SPECIAL PAPERS IN PALAEONTOLOGY NO. 78

GRAPTOLITES FROM THE UPPER ORDOVICIAN AND LOWER SILURIAN OF JORDAN

BY

DAVID K. LOYDELL

with 2 plates and 22 text-figures

THE PALAEONTOLOGICAL ASSOCIATION

LONDON

October 2007

CONTENTS

	Page
INTRODUCTION	5
PREVIOUS WORK	5
LOCALITY INFORMATION	6
BIOSTRATIGRAPHY	7
CHARACTERS	12
SYSTEMATIC PALAEONTOLOGY	14
Neodiplograptus	16
Normalograptus	29
Paraclimacograptus	47
Metaclimacograptus	48
Sudburigraptus	50
Cystograptus	51
Akidograptus	52
Parakidograptus	53
Rhaphidograptus	54
Dimorphograptus	55
Atavograptus	56
Huttagraptus	56
CONCLUSIONS	58
ACKNOWLEDGEMENTS	59
REFERENCES	59
APPENDIX	65

[Special Papers in Palaeontology 78, 2007, pp. 1–66]

Abstract Forty-two graptoloid graptolite species are described from the upper Hirnantian *persculptus* Biozone, lower Rhuddanian *ascensus-acuminatus* and *vesiculosus* biozones and Aeronian of Jordan. Nine species are left in open nomenclature; three are new: *Neodiplograptus lueningi*, *Ne.? opimus* and *Normalograptus bifurcatus*. Characters of use in defining uppermost Ordovician–lower Silurian biserial graptoloid taxa are discussed. In addition to dorso-ventral width and thecal spacing, the most important characters are defined by the nature of the growth of th1^1 and th1^2, which dictates the appearance of the proximal end. Other important characters include virgellar modifications and thecal morphology, particularly the inclination of thecal apertures and suprageni-cular walls. Sicular dimensions appear to be of little use in distinguishing between most taxa. Approximately half of the species described have been recorded previously in assem-blages from peri-Gondwanan Europe, to which the Jordanian assemblages are most similar. Occurrences of the North African biozonal index taxon *Neodiplograptus africanus* close to the *ascensus-acuminatus/vesiculosus* Biozone boundary in Jordan suggest that the base of the *africanus-tariti* Biozone may correlate with a level lower than previously thought. Records of *Normalograptus persculptus* in the Silurian all appear to be other taxa, with *N. persculptus* restricted to its upper Hirnantian biozone. Examination of the type material of *Normalograptus normalis* has revealed that most records of this taxon can be assigned to more slender species, such as *N. ajjeri*. The appendix lists the repositories and registration numbers of all the specimens examined in this study.

Key words: graptolite, Silurian, Ordovician, biostratigraphy, Jordan, Middle East, Gondwana.

RECENT interest in modelling the deposition of the organic-rich shales of latest Ordovician and early Silurian age in Jordan has led to the collection of a considerable number of graptolites in order to provide a biostrati-graphical framework for the models proposed (Armstrong *et al.* 2005, 2006; Lüning *et al.* 2005, 2006). Description of these specimens, together with earlier collections from Jordan housed in the Natural History Museum, London and the Sedgwick Museum, Cambridge, forms the basis of this paper. In total, 42 species are described, nine of which are left in open nomenclature. Three species, *Neodiplograptus lueningi*, *Ne.? opimus* and *Normalograptus bifurcatus*, are new.

The geological setting of the uppermost Ordovician and lower Silurian has been discussed at length by Lüning *et al.* (2005). All the graptolites described below are from the Mudawwara Formation, a shale-dominated unit deposited after the retreat of the late Ordovician ice from North Africa. Collections of graptolites are primarily from the Rhuddanian 'lower hot shale' (Lüning *et al.* 2005, fig. 2) and sub- and suprajacent strata.

PREVIOUS WORK

Although faunal lists appear in several works on the Lower Palaeozoic of Jordan, in very few papers are graptolites described or illustrated. Wolfart *et al.* (1968) described a low-diversity assemblage of Aeronian graptolites, including a new subspecies, *Climacograptus innotatus jordaniensis*, which is synonymized with *Paraclimacograptus libycus* (Desio, 1940) below. Armstrong *et al.* (2005, fig. 6) illustrated four specimens of *Normalograptus parvulus* (H. Lapworth, 1900) from a quarry section (Locality 7 below) and Lüning *et al.* (2005) three taxa [*Parakidograptus acuminatus* (Nicholson, 1867), *Neodiplograptus apographon* (Štorch, 1983a) and *Dimorphograptus confertus* (Nicholson, 1868a)] from core BG-14 (Locality 3 below).

Some papers (e.g. Powell *et al.* 1994, p. 305; Lüning *et al.* 2005, p. 1407) refer to biostratigraphical infor-mation in unpublished (and in many cases unobtainable via interlibrary loan) British Geological Survey, oil company and Jordanian Natural Resources Authority reports. Some of the graptolite material upon which this unpublished work was based is now housed in the Natural History Museum, London and in the Sedgwick Museum, Cambridge and has been examined for this paper.

Institutional abbreviations. The Jordanian specimens described herein are housed in the following institutions: British Geological Survey, Keyworth (BGS); Natural History Museum, London (NHM); Sedgwick Museum, Cambridge (SM); and the Geo-logical Directorate, Natural Resources Authority, Amman (GD-NRA). Other institutional abbreviations used are: BU, Lap-worth Museum, University of Birmingham; gr., Palaeontological

DAVID K. LOYDELL

School of Earth and Environmental Sciences, University of Portsmouth, Burnaby Road, Portsmouth PO1 3QL, UK; e-mail: david.loydell@port.ac.uk

 doi: 10.1111/j.1475-4983.2007.00712.x

Institute, Uppsala; GSC, Geological Survey of Canada, Ottawa; L, National Museum, Prague; LO, Department of Geology, University of Lund; PŠ, Geological Survey, Prague; UL, University of Lyon I, UMR Paléoenvironnements et Palaeobiosphère, Pôle Collection; UMPI, Palaeontological Institute, University of Modena; USNM, United States National Museum, Washington, DC.

LOCALITY INFORMATION

Material has been examined from 13 localities, as follows. Latitudes and longitudes were kindly provided by Sebastian Lüning and Howard Armstrong for material sent to me by them (indicated by SL and HA, respectively). For the Natural History Museum and Sedgwick Museum specimens, the locality information is that on the labels or in the drawers accompanying the specimens. Apart from the specimens from cores JF-1 and WS-6, all the material is from the Southern Desert area (Text-fig. 1). Institutional registration numbers are provided at the end of each locality description below.

1. Section through the Mudawwara Shale Formation at Batna El Ghoul 29°37.880'N, 35°54.597'E; graptolites from *c.* 6.25 m above base of measured section. Graptolite assemblage comprises *Normalograptus parvulus*, *Neodiplograptus parajanus*, *Ne. lanceolatus* and *Ne.* sp. 3, indicating the middle or upper *ascensus-acuminatus* Biozone (BGS FOR 5456–59) (SL).

2. Core BG-4, depths 4.9–16 m, from a shallow well in the Batna El Ghoul area. See Text-figure 2 for distribution of graptolites (GD-NRA 2402-3–2402-9) (HA).

3. Core BG-14 29°33.574'N, 35°58.528'E, depths 21.5–46.8 m, from a shallow kaolinite exploration well in the Batna El Ghoul area. Text-figure 3 shows the distribution of graptolites from the upper *ascensus-acuminatus* Biozone through to the *vesiculosus* Biozone (the highest sample, at 21.5 m, is of uncertain age) (BGS FOR 5363–5445, 5473; GD-NRA 8402-25, 8402-30–8402-37) (SL/HA).

4. Section, 6 m thick, through Mudawwara Shale Formation in abandoned kaolinite quarry, Jebel Batra (29°35.703'N, 35°57.538'E). Graptolite assemblage comprises *Normalograptus parvulus* and *N. ajjeri* (BGS FOR 5447–55) (SL).

5. Low hill *c.* 2 km west of Jebel Badra, near the main highway between Ma'an and Mudawwara; strata trenched during prospecting for clay; Mudawwara Formation. Graptolite assemblage comprises *Normalograptus persculptus*, *N. parvulus*, *N. bifurcatus* sp. nov., *N. ajjeri* and *Neodiplograptus* sp. 4, indicating the *persculptus* Biozone (NHM Q6336–41).

6. South-east of Sahl Essuwan, stated to be from the type locality of the 'Graptolite Sandstone' (Bender 1963, p. 251), from two horizons. The assemblage from the lower horizon comprises *Normalograptus parvulus* and *Neodiplograptus lanceolatus*, indicating the *ascensus-acuminatus* Biozone; the higher horizon contains only the Aeronian species *Paraclimacograptus libycus* (NHM QQ230–36).

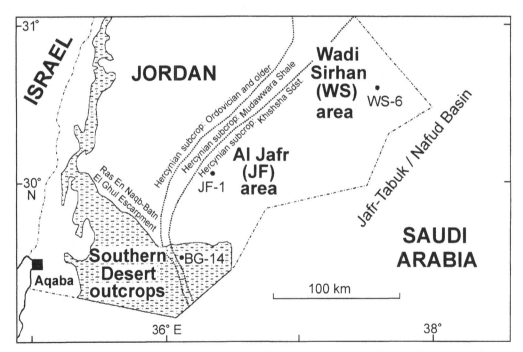

TEXT-FIG. 1. Map showing locations of boreholes BG-14, JF-1 and WS-6. The ornamented area is the outcrop of Palaeozoic sedimentary strata.

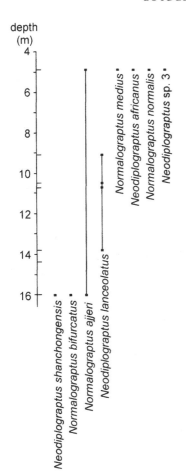

TEXT-FIG. 2. Graptolite distribution in core BG-4.

7. Working quarry at 29°42′00″N, 35°54′00″E. For distribution of graptolites, see Text-figure 4 (GD-NRA 18402/48–49, 51, 53–58); sample GD-NRA18402/54 is from the *vesiculosus* Biozone (HA).

8. Core WS-6, depths 1398.3–1399.8 m, from a deep hydrocarbon well in the Wadi Sirhan area, the graptolite assemblages from which were discussed by Lüning *et al.* (2005, p. 1407, fig. 1). See Text-figure 5 for distribution of graptolites, which indicate the lower part of the middle *ascensus-acuminatus* Biozone (SM A109351–52; BGS FOR 5464–72) (SL).

9. Section through the Mudawwara Shale Formation at Mudawwara (from 29°17.467′N, 36°07.741′E to 29°17.468′N, 36°07.420′E); one identifiable graptolite, *Normalograptus parvulus*, from *c.* 24.3 m above base of measured section (BGS FOR 5446) (SL).

10. Mudawwara Shale Formation in working kaolinite quarry at 29°37.593′N, 35°55.081′E. Graptolite assemblage comprises *Normalograptus rectangularis* and *Huttagraptus acinaces* (Törnquist, 1899), indicating the *vesiculosus* Biozone (BGS FOR 5460–63) (SL).

11. Sidewall core JF-1, in the Al-Jafr area. Identifiable graptolites occur at depths 3490 ft (*c.* 1064 m) (*Sudburi-graptus illustris*), 3494 ft (*c.* 1065 m) (*N. parvulus*) and 3497 ft (*c.* 1066 m) (*N. persculptus*). The presence of *N. persculptus* indicates the *persculptus* Biozone. Graptolites and palynomorphs from this core were discussed by Lüning *et al.* (2005, p. 1407) (SM A109362–67).

12. Shallow core ('#3') at 29°29′N, 35°53′E; graptolites from 12.5 to 13.0 m below wadi base. Two graptolites are present, *Normalograptus parvulus* and *Ne. apographon*, indicating the upper part of the *ascensus-acuminatus* Biozone (SM A109357–60).

13. Exposure of Mudawwara Formation at 29°29′N, 35°53′E with abundant graptolites. The species present are *Normalograptus parvulus*, *N. ajjeri*, *N. mirnyensis*, *N. rectangularis*, *N. normalis* and *Parakidograptus acuminatus*, indicating the upper part of the *ascensus-acuminatus* Biozone (SM A109353–56).

BIOSTRATIGRAPHY

With the exception of the Aeronian (Middle Llandovery) *Paraclimacograptus libycus*, all of the material described herein is of latest Ordovician (late Hirnantian) to early Llandovery (early Rhuddanian) age. This discussion is limited therefore to the biozonations used for the upper Hirnantian and lower Rhuddanian.

Upper Hirnantian

The *Normalograptus persculptus* Biozone is widely recognized as the uppermost Ordovician graptolite biozone, e.g. in Wales (see below), Bohemia (Štorch and Loydell 1996), Kazakhstan (Apollonov *et al.* 1980, 1988), northeast Russia (Koren' *et al.* 1988) and Australia (Vandenberg *et al.* 1984). In Sweden, however, Koren' *et al.* (2003) recognized an uppermost Hirnantian post-*persculptus* to pre-*ascensus* interval, characterized by the presence of *Normalograptus avitus* (Davies, 1929). *N. persculptus* appears to be restricted to its biozone; I can find no correctly identified Rhuddanian specimens in the published literature (see 'Systematic palaeontology' below). Thus its presence appears to be a clear indicator of a level within the eponymous biozone.

The *persculptus* Biozone was erected in mid Wales by Jones (1909). With the exception of Davies's (1929) work there has been little subsequent published research on the biozone in Wales. The impression gained (Davies 1929; Zalasiewicz and Tunnicliff 1994; my field work in the area) is that assemblages are of low diversity, being composed of *N. persculptus* and a small number of other *Normalograptus* species.

In some places *N. persculptus* has not been recorded from the upper Hirnantian and other taxa have been

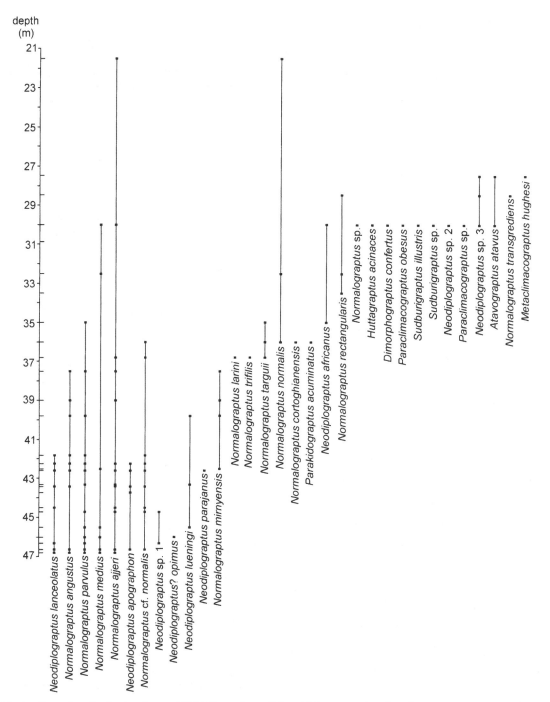

TEXT-FIG. 3. Graptolite distribution in core BG-14. Strata from 46.8 to 36.0 m are assigned to the upper *ascensus-acuminatus* Biozone and from 30.0 to 27.5 m to the *vesiculosus* Biozone.

considered to be indicative of the *persculptus* Biozone. At Dob's Linn, for example, the GSSP for the base of the Silurian, Williams's (1983) extensive collecting did not produce any specimens of the species. Melchin *et al.* (2003, p. 101) reported *N. persculptus* from Dob's Linn, but from the '*extraordinarius* Band', lending support to Štorch and Loydell's (1996) suggestion that the *extraordinarius* Biozone should be relegated to the rank of a sub-zone representing the lower part of the *persculptus*

Biozone. Similarly, in some Chinese sections, *N. persculptus* is not recorded (Chen and Melchin 2005*a*, text-fig. 2). It is interesting to note that where *N. persculptus* is recorded from the most intensively studied Chinese sections (Wangjiawan and Fengxiang; Chen *et al.* 2005*a*) the species is restricted to the lower, or lower and middle parts of its biozone, lending support to the post-*persculptus* to pre-*ascensus* interval proposed by Koren' *et al.* (2003).

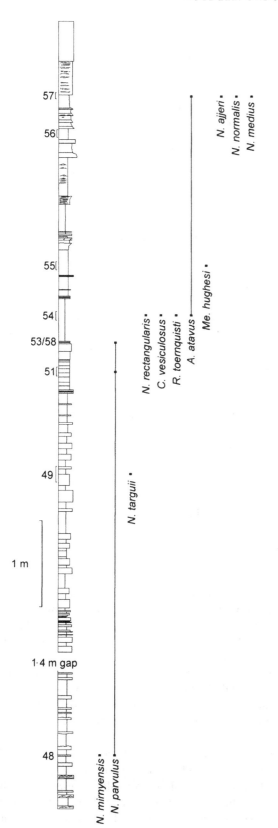

TEXT-FIG. 4. Graptolite distribution in working quarry at 29°42′00″N, 35°54′00″E (Locality 7). Log kindly provided by Howard Armstrong.

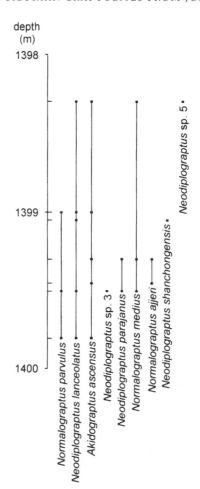

TEXT-FIG. 5. Graptolite distribution in the lower part of the middle *ascensus-acuminatus* Biozone of core WS-6.

In Jordan *N. persculptus* and thus its biozone have been identified. Elsewhere on Gondwana, however, e.g. southern Algeria and southern Libya, the graptolite biozonation of the upper Hirnantian and the Rhuddanian has been based on endemic biserial taxa (e.g. Legrand 2003, fig. 2). Neither *N. persculptus*, nor any of the taxa used elsewhere to recognize the biozone, has been identified in these parts of North Africa, and correlations with sections elsewhere in the world are usually described as 'tentative' or 'approximate' (e.g. Legrand 2003). Legrand (2001, fig. 8f), however, illustrated a specimen which he identified as ' "*Glyptograptus*" ('*Glyptograptus*'), e.g. *persculptus*', from close to the Algeria-Niger border, which I consider to be *N. persculptus*, which indicates that this biostratigraphically important species occurs in North Africa, albeit very rarely.

Lower Rhuddanian

A basal Rhuddanian *ascensus-acuminatus* Biozone is widely recognized, although sometimes it is referred to simply as the *acuminatus* Biozone. In Poland (Teller

1969), Thuringia (Schauer 1971) and in the most recent work on Bohemia (Štorch 2006) the interval is split into two biozones, a lower *ascensus* Biozone and a higher *acuminatus* Biozone (Schauer 1971). The latter division reflects the fact that the first appearance of *Akidograptus ascensus* precedes that of *Parakidograptus acuminatus*; there is, however, considerable overlap in their stratigraphical ranges (Text-fig. 6). It has long been recognized that subdivision of the *ascensus-acuminatus* Biozone is possible. Rickards (1988, p. 346), for example, commented on *Normalograptus trifilis* Manck, 1923 characterizing the middle part of the biozone and *N. rectangularis* (M°Coy, 1850) appearing in the upper part of the biozone.

The *ascensus-acuminatus* Biozone is succeeded by either the *vesiculosus* Biozone or, where *Cystograptus vesiculosus* is rare or absent (e.g. Wales), by the *atavus* and then *acinaces* biozones. Localities and regions that have been the subject of recent study of the lower Rhuddanian are discussed below, as is the biozonation in Jordan.

Dob's Linn, Moffat, southern Scotland. The GSSP for the base of the Silurian is the current focus of detailed study, with a number of abstracts published (e.g. Melchin and Williams 2000; Melchin 2003). It is clear from these abstracts and examination of Toghill's collections in the NHM that the diversity of lower Rhuddanian graptolites shown by Toghill (1968) and Williams (1983) is a considerable underestimate. Melchin (2003) recorded several taxa from Dob's Linn that have proved biostratigraphically useful in Uzbekistan (Koren' and Melchin 2000) and

China (e.g. Chen and Lin 1978). The presence of these and previously recorded taxa known to have a restricted range (e.g. *N. trifilis*; see Štorch 1996, fig. 2; Text-fig. 6) suggests that a very fine-scale biostratigraphical subdivision and precise correlation of the lower Rhuddanian of Dob's Linn will in the future be possible. In comparison with peri-Gondwanan Europe (and Jordan) I have found that *Neodiplograptus* is rare at Dob's Linn.

Peri-Gondwanan Europe. Štorch (1996, p. 177) used this term for those parts of western, central and southern Europe that during the early Palaeozoic 'were linked tightly to the African part of the north Gondwana margin'. Štorch (1996) provided a very useful review of peri-Gondwanan graptoloid assemblages from the *ascensus-acuminatus* Biozone, noting the 'rather uniform' nature of the assemblages throughout the region. He recognized an informal three-fold subdivision of the biozone (reproduced as Text-figure 6), which is generally rather condensed, e.g. 0.5–1.0 m thick in Germany, up to 2.5 m thick in Bohemia; only biserial graptolites have been recorded from the *ascensus-acuminatus* Biozone in this region. In Bohemia the *ascensus-acuminatus* Biozone is succeeded by the *vesiculosus* Biozone which yields, in addition to the index species, the earliest 'monograptids' and the first (and only) *Dimorphograptus*, *D. confertus*, to be recorded from Bohemia (Štorch 1994).

Kurama Range, east Uzbekistan. Koren' and Melchin (2000) described the lowermost Rhuddanian graptolites of this region. Within the Mashrab Valley, three local subzones of the *ascensus-acuminatus* Biozone were recognized, in ascending order: *Normalograptus lubricus*, *Akidograptus cuneatus* and *Hirsutograptus sinitzini*. The transition from the last into higher stratigraphical levels is not preserved in the region because of faulting. As noted below, the specimens assigned by Koren' and Melchin (2000) to *P. acuminatus* are not this species and it is very likely that their subzones are divisions of only the lower part of the *ascensus-acuminatus* Biozone. This is supported by the presence of *A. ascensus* in the highest, *sinitzini*, subzone (Koren' and Melchin 2000, table 1): elsewhere it disappears some distance below the top of the *ascensus-acuminatus* Biozone (e.g. Štorch 1996, fig. 2; Text-fig. 6).

Poland. Masiak *et al.* (2003) described the graptolite assemblages through the *ascensus-acuminatus* and *vesiculosus* biozones of the Holy Cross Mountains. All three biozonal indicators were present, together with reasonably diverse assemblages, comparable with those described elsewhere in Europe.

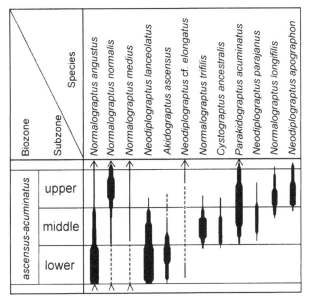

TEXT-FIG. 6. Divisions of the *ascensus-acuminatus* Biozone of peri-Gondwanan Europe (after Štorch 1996, fig. 2).

Wales. Zalasiewicz and Tunnicliff (1994, p. 701) suggested that the 'gross composition of [graptolite]

assemblages can be used, cautiously, to give some evidence of age'. In their collections from Wales, virtually all Rhuddanian collections from above the *ascensus-acuminatus* Biozone contained 'monograptids' and/or *Rhaphidograptus*. Many of these collections (27 out of 48) were very small (5–9 specimens), suggesting that these two components of graptolite assemblages were sufficiently common to be encountered even in such small collections. Many collections from lower horizons (*ascensus-acuminatus* and *persculptus* biozones) contained nothing but *Normalograptus*.

The dominance of *Normalograptus* in the *ascensus-acuminatus* Biozone is much less apparent in the Jordanian assemblages, in large part due to the much greater abundance of *Neodiplograptus* here than in Wales. For example, in the BG-14 core, only four samples (of five specimens or more) out of 20 are exclusively *Normalograptus*. Of these, three are from the upper part of the *ascensus-acuminatus* Biozone; the other is the highest sample (21.5 m), the age of which is uncertain.

The four Jordanian samples (of five specimens or more) from the *vesiculosus* Biozone all contain *Atavograptus atavus* (Jones, 1909), i.e. a 'monograptid', and one contains *Rhaphidograptus*. These limited Jordanian data thus agree with Zalasiewicz and Tunnicliff's (1994) observation that virtually all Rhuddanian collections from above the *ascensus-acuminatus* Biozone contain 'monograptids' and/or *Rhaphidograptus*.

Southern Sahara. Legrand (2003, fig. 2) summarized the biozonation for the southern Saharan region of North Africa. As for the upper Hirnantian, the Rhuddanian index taxa were considered to be endemic, with the exception of *Neodiplograptus fezzanensis* (Desio, 1940), which is also known from the *cyphus* and lower *triangulatus* biozones of Bohemia (Štorch 1983a). Some of these 'endemic' taxa have been encountered in the Jordanian collections (see below), associated with geographically more widespread species, offering the opportunity to tie some of Legrand's biozones more precisely to the 'standard' graptolite biozonation.

Mauritania. The collections of Underwood *et al.* (1998) lack the biozonal index taxa for the lower Rhuddanian, but do include some widespread, biostratigraphically useful species, e.g. *N. rectangularis* and *Metaclimacograptus hughesi*. They showed the lowermost Rhuddanian to be dominated by *Normalograptus*, paralleling the situation in Wales (see above).

Jordan. The presence of the biozonal indices *A. ascensus*, *P. acuminatus* and *C. vesiculosus* enables recognition of the 'standard' lower Rhuddanian graptolite biozones in Jordan. In addition, the presence of the biostratigraphically useful species *Neodiplograptus apographon* enables

recognition of the upper *ascensus-acuminatus* Biozone of Štorch (1996; Text-fig. 6).

When my work on the Jordanian graptolites commenced, the expectation was that the assemblages would be most similar to those described from North Africa, because of their proximity to North Africa during the Silurian and the description by Wolfart *et al.* (1968) of *Climacograptus innotatus jordaniensis*, a junior synonym of the endemic North African species *Paraclimacograptus libycus*. It became apparent, however, that the Rhuddanian assemblages are similar to those of peri-Gondwanan Europe, but with a few 'North African' species also present: *Normalograptus targuii* Legrand, 2001, *Neodiplograptus africanus* (Legrand, 1970) and *Ne.* sp. 3 (as described below). The two *Neodiplograptus* species are index species in Legrand's (2003, fig. 2) biozonation. *Ne.* sp. 3 was considered by Legrand to occur at a level equivalent to the lower *vesiculosus* Biozone. Although the species occurs in Jordan in the *vesiculosus* Biozone of the BG-14 core, it also occurs stratigraphically much lower, with *Akidograptus ascensus* in the WS-6 core, indicating a significantly longer stratigraphical range than indicated by Legrand. *Ne.* sp. 3 is probably not therefore useful in high-resolution biostratigraphy. *Ne. africanus* occurs in the BG-14 core both in the *vesiculosus* Biozone and in an unzoned interval only 1 m above a level bearing *P. acuminatus*. Legrand (2003, fig. 2) suggested that the '*Nd. africanus* and "*Gl.*" *tariti* Zone' correlated with the middle *vesiculosus* to lower *cyphus* biozones. It would seem that in Jordan at least the first specimens of *Ne. africanus* occur a little lower, close to the *ascensus-acuminatus/vesiculosus* Biozone boundary.

The *ascensus-acuminatus* Biozone clearly attains a considerable thickness in Jordan. In the BG-14 core the upper part of the biozone has a minimum thickness of 10.8 m, compared with a maximum thickness of 2.5 m for the entire biozone in Bohemia. It would seem likely that only parts of the *ascensus-acuminatus* Biozone have thus far been sampled from Jordan. For instance, the absence in any sample of abundant *Normalograptus trifilis*, a species common in the middle *ascensus-acuminatus* Biozone of peri-Gondwanan Europe and elsewhere (e.g. Dob's Linn), suggests that this level has not been examined, except probably the lowermost part in core WS-6. Similarly, although the lower part of the *ascensus-acuminatus* Biozone may be represented by some small collections of *Normalograptus parvulus* and/or *Neodiplograptus lanceolatus*, these could equally be from higher in the biozone. It would seem likely therefore that the true diversity of late Hirnantian–early Rhuddanian graptolites in Jordan, already high at more than 40 species, will prove significantly higher once the entire upper Hirnantian–lower Rhuddanian has been sampled.

CHARACTERS

With the exception of four well-known uniserial or uni-biserial taxa, all of the graptolites examined were biserials. This discussion is limited to consideration of characters useful in the identification of Hirnantian and lower Rhuddanian biserial graptolites at species level, with emphasis on those characters generally visible in non-isolated material. Some of the features measured in the Jordanian material are illustrated in Text-figure 7. Silurian biserial graptolite genera have been discussed at length by Melchin (1998) and his generic definitions are those followed here, with the exception of *Paraclimacograptus*, which was revised by Russel *et al.* (2000).

One of the problems with the identification of biserial graptolites, and it seems particularly those from high palaeolatitudes, is that they show considerable intra-specific variation. Legrand (e.g. 1970, 1977, 1999) has documented this in many species of *Normalograptus* and *Neodiplograptus* from Algeria, providing tables and graphs of measurements and illustrations of the graptolites to demonstrate its extent.

Sicula

Most published descriptions of graptolite species include details of sicular length, apertural width and the level within the rhabdosome attained by its apex. These measurements are often more easily made in flattened material as the sicula may be 'pressed through' the rhabdosome; in relief specimens the sicular apex is often obscured by the growth of the early thecae. It is very rarely possible to distinguish between prosicula and metasicula in non-isolated material. Unless exceptionally long (as in *Cystograptus* Hundt, 1942) or short (e.g. *Normalograptus legrandi* Koren' and Rickards, 2004), sicular dimensions appear generally too similar to be of value in distinguishing between taxa. The sicular lengths of Jordanian *Normalograptus* are compared in Text-figure 8. Although the sicula of the Jordanian specimens of *Neodiplograptus parajanus* Štorch, 1983a is longer than in any other Jordanian biserial examined, other than *C. vesiculosus* (Nicholson, 1868b), Štorch and Serpagli (1993, p. 18) recorded a greater range of sicular length in their new species *Ne. lanceolatus* than seen in Jordanian material of this species, up to 2.5 mm, suggesting that the long sicula of *Ne. parajanus* may not be a diagnostic character. Štorch (1983b) suggested that in *Akidograptus ascensus* sicular length may decrease through time. The position of the apex of the sicula reflects a combination of sicular length, length of downward-growing portion of $th1^1$, and proximal thecal spacing.

Virgella and other sicular spines

Virgellar length appears to be highly variable within species, although in some taxa, e.g. *Normalograptus parvulus*, only short virgellae are known. The virgella is likely to be damaged biostratinomically or during collection and preparation of graptolite specimens, and it can also sometimes be difficult to determine whether a short virgella is complete or broken. Virgellar length seems to be of limited value in specific determination.

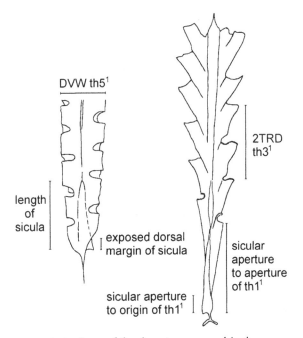

TEXT-FIG. 7. Some of the characters measured in the Jordanian biserial graptolite taxa.

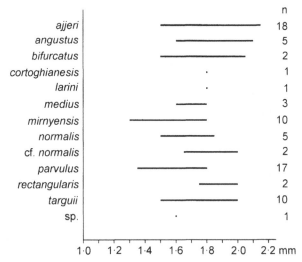

TEXT-FIG. 8. A comparison of sicular length in Jordanian *Normalograptus* species.

Virgellar breadth is in some cases a diagnostic character. *Normalograptus balticus* (Pedersen, 1922), for example, is recognized by its extremely robust virgella that appears to merge with the proximal end of the thecate portion of the rhabdosome (Bjerreskov 1975, p. 25). In most taxa, however, the virgella is a narrow spine.

Virgellar bifurcation or branching is of considerable significance in defining and identifying taxa. Several latest Ordovician and early Silurian biserials exhibit modified virgellae; examples are illustrated by Chen and Lin (1978, e.g. fig. 4), Štorch (1996, fig. 4) and Chen *et al.* (2005*a*, e.g. text-fig. 8). In *N. bifurcatus* sp. nov. the distance between the sicular aperture and virgellar bifurcation is very variable. Complex structures are known in some taxa, e.g. *Normalograptus* sp. B (Štorch 1996, fig. 3f), and in *Parakidograptus* and *Akidograptus* the branching is similar to that of the ancorae that characterize *Petalolithus* Suess, 1851, *Pseudorthograptus* Legrand, 1987 and the Silurian retiolitids.

Additional sicular spines are present in *Hirsutograptus* Koren' and Rickards, 1996, in which they are accompanied by thecal apertural spines. A bifurcating antivirgellar spine is present in *Normalograptus bifurcatus* sp. nov., which has non-spinose thecae. These additional sicular spines are important in defining these species.

Growth of early thecae

With the exception of *Akidograptus* Davies, 1929 and *Parakidograptus* Li and Ge, 1981, all taxa exhibit significant initial downward growth of th1[1], before it turns and grows upwards. In all cases this downward growth extends from the metasicular origin of th1[1] to below the sicular aperture. In most of the material described here, the position of the origin of th1[1] is not visible, so the measurements provided are of the downward growth of th1[1] below the sicular aperture. In *Paraclimacograptus libycus*, this appears greatest (Text-fig. 20C–H), but the 0.3–0.35 mm distances recorded in the Jordanian material of this species are exceeded by the highest figure recorded (0.4 mm) in *Normalograptus parvulus*, *N. rectangularis* and *N. targuii*. Within individual species there appears to be considerable variation in this character, e.g. in all three of the above taxa the range is 0.1–0.4 mm. Of considerably more significance in species recognition are the inclination and curvature of the ventral walls of th1[1] and th1[2], because it is these characters that determine to a large degree the appearance of the proximal end.

In *P. acuminatus* it has been suggested that at higher stratigraphical levels the species has a less protracted proximal end (Štorch 1983*b*, p. 298). Quantification of this character is possible by measuring the distance from the sicular aperture to the aperture of th1[1] (Text-fig. 7) and the 2TRD (see below) of proximal thecae.

Dorso-ventral width (DVW) and thecal spacing

These two characters are the most widely recorded in graptolite descriptions and they have the considerable advantage of being measurable in all but the most poorly preserved material. Differences in DVW and/or thecal spacing, usually now quoted as two thecae repeat distances (2TRD; Howe 1983) at specified thecae, are very important in distinguishing, in tectonically undeformed material, between many otherwise morphologically similar taxa. For example, *Normalograptus angustus* and *N. mirnyensis* are distinguished largely on the basis of the wider thecal spacing of the former; *N. parvulus* and *N. persculptus* are distinguished by the higher distal DVW and/or 2TRD of the latter (Text-fig. 9).

Thecae

In *Normalograptus*, *Sudburigraptus*, *Paraclimacograptus* and *Metaclimacograptus* the thecae are of essentially uniform morphology along the length of the rhabdosome. In *Neodiplograptus*, however, proximal thecae have sharp genicula; distally the thecal morphology changes either rapidly or gradually to glyptograptid and/or orthograptid.

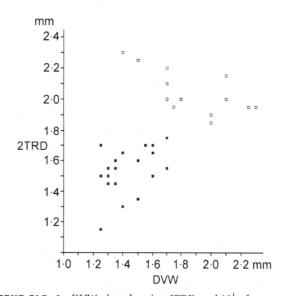

TEXT-FIG. 9. DVW plotted against 2TRD at th10[1] of specimens of *Normalograptus parvulus* (black squares) and *N. persculptus* (open squares). Measurements of *N. parvulus* are from specimens from Jordan, the United States (Loydell *et al.* 2002) and Dob's Linn (Williams 1983). Measurements of *N. persculptus* are from specimens from Wales and Bohemia (Štorch and Loydell 1996) and Kazakhstan (Koren' *et al.* 1980).

The orientation of supragenicular walls is an important character. It appears to be consistent within taxa and the identification of *Normalograptus targuii* and *N. parvulus* is based in part upon their inclined supragenicular walls. Preservation mode influences the angle of inclination: in full relief material of *N. parvulus* (such as illustrated by Armstrong *et al.* 2005, fig. 6) the supragenicular walls are inclined at a much lower angle to the rhabdosome axis than in flattened specimens (compare Text-fig. 18I with 18J).

The length of the supragenicular wall may also be of significance, particularly when compared with the length of the aperture. In *Paraclimacograptus libycus*, for example, the supragenicular walls are short, whilst in *Normalograptus melchini* they are long.

Depth of thecal excavations is of significance in the recognition of some species; for example, *Normalograptus medius* has deep excavations and in *N. melchini* they are very shallow. However, in oblique views, which are common in *Normalograptus* because of the almost cylindrical nature of the rhabdosome, this character may be difficult to assess accurately.

The orientation of thecal apertures appears to be a significant character in the identification of some *Neodiplograptus* species; for example, *Ne. modestus* has horizontal apertures distally, whereas in *Ne. lanceolatus* they are everted.

Thecal spines occur in *Hirsutograptus* Koren' and Rickards, 1996, a genus not recorded from Jordan. The only thecal outgrowths seen in Jordanian material are the genicular flanges of *Paraclimacograptus* Přibyl, 1947.

The inclination of the thecae to the rhabdosomal axis appears to be useful in the recognition of some taxa, e.g. Rhuddanian specimens assigned to *N. persculptus* by Zalasiewicz and Tunnicliff (1994) are rejected as belonging to this species based on the inclination of the proximal thecae (see below). The amount of thecal overlap and the course of interthecal septa are characteristic in some taxa, e.g. *N. persculptus* in which the interthecal septa exhibit sigmoidal curvature. As with the median septum (see below) these characters may be difficult to assess in less well-preserved material and in specimens with thick periderm.

Nema

With the exception of *Cystograptus*, in which a nematularium is developed (Jones and Rickards 1967;

Urbanek *et al.* 1982), the nemata of Hirnantian and Rhuddanian biserials are generally of simple form. Štorch (1996, fig. 4.1c) illustrated outgrowths from the nema of a specimen identified as '*Glyptograptus*? *avitus* Davies'.

Median septum

The absence or presence of a median septum, whether complete or partial, is a character of significance in the definition of several genera of biserial graptolites (Melchin 1998). The level within the rhabdosome of the point of origin of the median septum is important in the recognition of some *Normalograptus* species, e.g. *N. transgrediens* (Wærn, 1948). Several authors (e.g. Davies 1929; Wærn 1948) recognized that the origin of the median septum is successively delayed in stratigraphically younger material of several species.

In flattened material it is often difficult to see whether a median septum is present, particularly in less well-preserved material and in taxa in which the periderm is thick. Fortunately, in some of the Jordanian material, particularly of *Normalograptus*, the junctions of the median septum with the obverse and reverse sides of the rhabdosome can be seen. In many specimens, however, it is difficult to recognize either the presence of a median septum or the level of its origin, thus limiting the potential value of these characters in identification. In relief material, it is generally impossible to determine from an obverse view whether a median septum is complete or partial.

SYSTEMATIC PALAEONTOLOGY

All specimens are preserved flattened unless stated otherwise. In many cases part of the periderm has flaked off the bedding surface, making many specimens unsuitable for photography. All line drawings and photographs are at ×10 magnification to aid comparison between species. A few specimens were too long to figure in their entirety at ×10 and have been included in Text-figure 12 at ×5.

Because most biserial taxa cannot be identified unless the proximal end is present, the total number of graptolites examined was far in excess of the number that could be identified.

TEXT-FIG. 10. A–B, *Neodiplograptus africanus* (Legrand, 1970), core BG-14. A, BGS FOR 5388, 35.0 m, upper *ascensus-acuminatus* or *vesiculosus* Biozone. B, BGS FOR 5369l, 30.0 m, *vesiculosus* Biozone. C–D, F–G, *Neodiplograptus apographon* (Štorch, 1983*a*), core BG-14, 43.4 m, upper *ascensus-acuminatus* Biozone. C, BGS FOR 5359c. D, BGS FOR 5359i. F, BGS FOR 5359a (figured by Lüning *et al.* 2005, fig. 15b). G, BGS FOR 5359b. E, I–J, *Neodiplograptus lanceolatus* Štorch and Serpagli, 1993. E, BGS FOR 5471b, core WS-6, 1399.5 m, middle *ascensus-acuminatus* Biozone. I, BGS FOR 5466a, core WS-6, 1399.05 m, middle *ascensus-acuminatus* Biozone. J, BGS FOR 5359d, core BG-14, 43.4 m, upper *ascensus-acuminatus* Biozone. H, *Paraclimacograptus* sp., BGS FOR 5371a, core BG-14, 30.0 m, *vesiculosus* Biozone. All figures × 10.

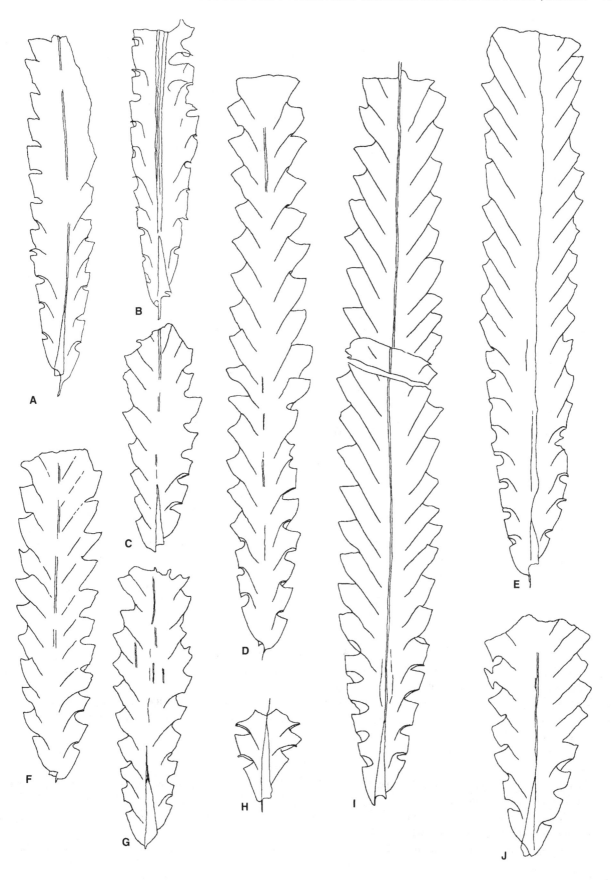

Genus NEODIPLOGRAPTUS Legrand, 1987; emend.
Melchin, 1998

Neodiplograptus africanus (Legrand, 1970)
Text-figure 10A–B

*. 1970 *Diplograptus africanus africanus* nov. sp. Legrand,
p. 7, figs 2-011a–b, 2-013a–n.
2002 *Neodiplograptus africanus africanus* (Legrand);
Legrand, pl. 12, fig. 4; text-fig. 5d–f.
v. 2003 *Neodiplograptus africanus* (Legrand); Lüning
et al., fig. 6.
2004 *Neodiplograptus africanus* (Legrand); Štorch and
Massa, fig. 3.6–3.8.

Material. Five specimens: two from a depth of 30.0 m, one
from 35.0 m, core BG-14; two from a depth of 4.9 m, core
BG-4.

Description. The sicula is 1.6–1.7 mm long. The sicular apex
attains the level of the aperture of th2^1 or th2^2. Th1^1 grows
downwards for 0.15–0.25 mm below the sicular aperture. Proximal thecae are of climacograptid form, although with gently
inclined supragenicular walls, at up to 10 degrees to the rhabdosome axis. Apertural excavations are horizontal and generally are
shallow. Distally, the angle of inclination of the supragenicular
walls increases to 15 degrees, although the geniculum remains
sharp. Measurements of DVW and 2TRD are given below. There
is a complete median septum.

DVW and 2TRD measurements (in mm)

th	1^1	2^1	3^1	5^1	10^1	15^1
DVW	0.9–1.0	1.1–1.3	1.3–1.5	1.5–1.8	1.75–1.8	2.0
2TRD		1.25–1.45	1.25–1.5	1.3–1.65	1.4–1.65	1.7

(n = 5, except for th10^1 measurements, where n = 3, and th15^1
measurements, where n = 1).

Remarks. Legrand (1970) observed considerable intraspe-
cific variation in *Ne. africanus*. The Jordanian specimens
lie toward the narrow end of the range of variation.

Stratigraphical range. Ne. africanus is an index species of the
North African '*Nd. africanus* and "*Gl.*" *tariti* Zone' of Legrand
(2003). It occurs in the BG-14 core both in the *vesiculosus* Biozone and in an unzoned interval only 1 m above a level bearing
P. acuminatus.

Neodiplograptus apographon (Štorch, 1983a)
Plate 1, figure 1; Plate 2, figure 2; Text-figure 10C–D, F–G

? 1965 *Diplograptus (Diplograptus?)* sp. A; Stein, p. 178,
text-fig. 24a.
p 1979 *Diplograptus modestus* cf. *modestus* Lapworth;
Jaeger and Robardet, pl. 2, fig. 3 (*non* fig. 6,
text-fig. 9b–c).
*. 1983a *Diplograptus diminutus apographon* ssp. n.
Štorch, p. 163, pl. 1, figs 1–2, text-fig. 2A–D.
1993 *Neodiplograptus diminutus apographon* (Štorch,
1983); Štorch and Serpagli, p. 17, pl. 1, fig. 1;
text-fig. 5H.
1996 *Neodiplograptus diminutus apographon* (Štorch);
Štorch, fig. 4.4a.
? 1999 '*Neodiplograptus*' sp. 2; Maletz, p. 350, fig. 2/12.
v. 2005 *Neodiplograptus diminutus apographon* (Štorch,
1983); Lüning *et al.*, fig. 15a.

Holotype. PŠ 72/1, figured by Štorch (1983a, pl. 1, fig. 1,
text-fig. 2C), from the Želkovice Formation of Praha-Řepy;
Parakidograptus acuminatus Biozone.

Material. 17 flattened specimens; 16 from core BG-14, depths
42.25, 42.6, 43.4 and 46.62 m; upper *ascensus-acuminatus*
Biozone; one from Locality 12.

EXPLANATION OF PLATE 1

Fig. 1. *Neodiplograptus apographon* (Štorch, 1983a). BGS FOR 5359a, specimen figured by Lüning *et al.* (2005, fig. 15b), upper
ascensus-acuminatus Biozone, core BG-14, 43.4 m.

Fig. 2. *Neodiplograptus? opimus* sp. nov. Holotype, GD-NRA 8402-31d, upper *ascensus-acuminatus* Biozone, core BG-14, 46.0 m.

Fig. 3. *Normalograptus angustus* (Perner, 1895). Holotype, L 27507, upper part of the Králův Dvůr Formation, Králův Dvůr, Bohemia.
Image kindly supplied by Petr Štorch.

Figs. 4, 7. *Normalograptus normalis* (Lapworth, 1877). 4, BGS FOR 5378a, upper *ascensus-acuminatus* or *vesiculosus* Biozone, core
BG-14, 32.5 m. 7, BGS FOR 5400a, upper *ascensus-acuminatus* Biozone, core BG-14, 36.0 m.

Fig. 5. *Normalograptus bifurcatus* sp. nov. Holotype, NHM Q6337(3), proximal end, showing clearly the separate origins of the virgella
and antivirgellar spine, *persculptus* Biozone, Locality 5.

Fig. 6. *Parakidograptus acuminatus* (Nicholson, 1867), BGS FOR 5360a, specimen figured by Lüning *et al.* (2005, fig. 15a), upper
ascensus-acuminatus Biozone, core BG-14, 36.0 m.

Fig. 8. *Normalograptus rectangularis* (McCoy, 1850), BGS FOR 5378b, upper *ascensus-acuminatus* or *vesiculosus* Biozone, core BG-14,
32.5 m.

All figures × 10.

PLATE 1

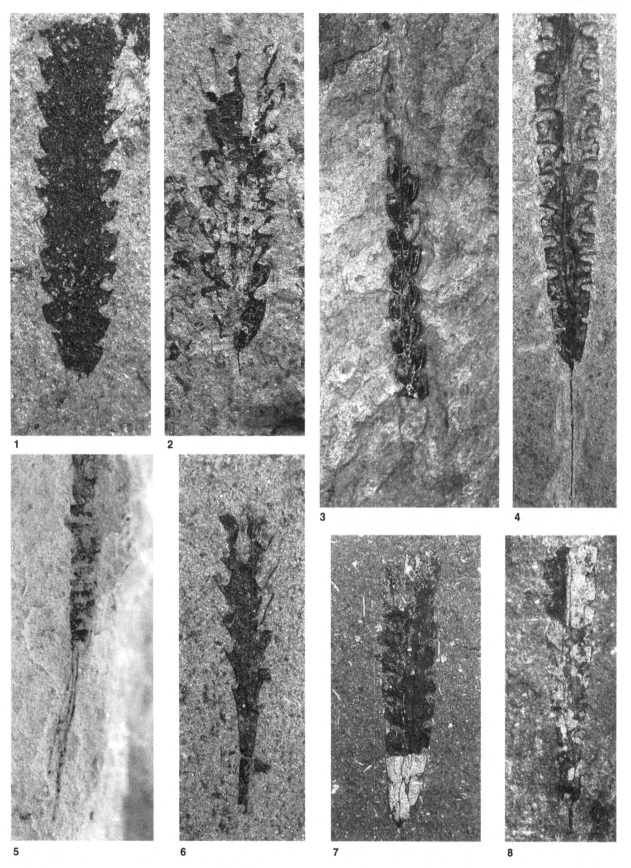

Description. The rhabdosome is generally short, the longest specimen being 15 mm long. The proximal end is conspicuously rounded. The sicula is 1.55–1.65 mm long ($n = 4$), 0.3 mm wide at its aperture. Its apex reaches to the apertures of the second thecal pair or just above. The aperture bears a short virgella, up to 0.25 mm long. Th1[1] grows downwards for 0.15–0.25 mm below the sicular aperture ($n = 4$). The length of the exposed part of the dorsal sicular margin is 0–0.35 mm. Proximal thecae have an angular geniculum; distally this geniculum becomes much less pronounced, the thecae becoming straight or very nearly so. Thecae are inclined to the rhabdosome axis at 20–30 degrees proximally, 30–40 degrees distally. Thecal apertures are everted, more so distally than proximally. Proximal DVW and 2TRD measurements are given below. There is a complete median septum.

DVW and 2TRD measurements (in mm)

th	1[1]	2[1]	3[1]	5[1]	10[1]
DVW	0.95–1.25	1.25–1.75	1.4–1.85	1.7–2.1	1.8–2.15
2TRD		1.25–1.5	1.2–1.55	1.35–1.65	1.55–1.75

($n = 12$–14 for measurements at th1[1]–5[1]; $n = 3$ for measurements at th10[1]).

Remarks. This Jordanian material is identical to that described by Štorch (1983a) from Bohemia. *Neodiplograptus apographon* differs from *Ne. diminutus* (Elles and Wood, 1907) in being broader and in having more widely spaced thecae. Maletz's (1999) '*Ne.*' sp. 2 is very similar to *Ne. apographon* in terms of DVW, but his figured specimen has more closely spaced thecae (2TRD at th2[1] = 0.95 mm), however, than recorded in *Ne. apographon*.

Stratigraphical range. Štorch (1983a, p. 164) recorded *Ne. apographon* from the middle and upper *acuminatus* Biozone. The Jordanian material described herein is from the same stratigraphical level. Štorch and Serpagli's (1993) specimens were from the upper *acuminatus* Biozone. Maletz's (1999) '*Ne.*' sp. 2, placed only questionably in synonymy, is from a lower stratigraphical level, within the upper part of the range of *A. ascensus*, but below the first *P. acuminatus*.

Neodiplograptus lanceolatus Štorch and Serpagli, 1993
Plate 2, figure 6; Text-figures 10E, I–J, 11A–B, E, G

p	1979	*Diplograptus modestus* cf. *modestus* Lapworth; Jaeger and Robardet, pl. 2, fig. 6 (*non* fig. 3); text-fig. 9b–c.
	1962	*Diplograptus modestus modestus* Lapw.; Tomczyk, pl. 4, fig. 5; pl. 7, fig. 6.
	1983a	*Diplograptus modestus modestus* Lapworth, 1876; Štorch, p. 162, pl. 1, figs 3, 5–6; text-fig. 2E–I.
?	1984	*Diplograptus modestus* Lapworth; Li, p. 342, pl. 5, fig. 9; pl. 12, figs 7–9; pl. 13, figs 1–4.
	1986	*Diplograptus modestus* Lapworth; Štorch, pl. 5, fig. 2.
?	1990	*Diplograptus elacatus* Yu *et al.*; Fang *et al.*, p. 47, pl. 1, figs 1–6, 9–10, 13; pl. 2, fig. 1; pl. 3, figs 1, 6; pl. 4, fig. 5.
	1990	*Diplograptus modestus* Lapworth; Gnoli *et al.*, text-fig. 4e.
*.	1993	*Neodiplograptus lanceolatus* n. sp.; Štorch and Serpagli, p. 18, pl. 2, figs 4, 7–8; text-fig. 6A–C. [with additional synonymy].
	1996	*Neodiplograptus lanceolatus* Štorch and Serpagli, 1993; Štorch, fig. 4.4d.
p	1999	'*Neodiplograptus*' *lanceolatus* (Štorch and Serpagli, 1993); Maletz, p. 349, fig. 3/3 (?fig. 2/9, *non* figs 3/1, 3/7, 4/7, 4/10).

Holotype. UMPI 22405a, figured by Štorch and Serpagli (1993, pl. 2, fig. 8; text-fig. 6A), from the Genna Muxerru Formation of Monte Cortoghiana Becciu, Sardinia; *P. acuminatus* Biozone.

Material. 140 specimens: 65 from core BG-14, depths 41.85, 42.25, 42.6, 43.4, 44.5, 46.3, 46.62 and 46.8 m; upper *ascensus-acuminatus* Biozone; 28 from core WS-6, depths 1398.3, 1399.0, 1399.05, 1399.5 and 1399.8 m; 40 from core BG-4,

---

EXPLANATION OF PLATE 2

Fig. 1. *Normalograptus larini* Koren' and Melchin, 2000. BGS FOR 5407b, upper *ascensus-acuminatus* Biozone, core BG-14, 36.8 m.

Fig. 2. *Neodiplograptus apographon* (Štorch, 1983a). SM A109359, upper *ascensus-acuminatus* Biozone, Locality 12.

Fig. 3. *Normalograptus cortoghianensis* (Štorch and Serpagli, 1993). BGS FOR 5400b, upper *ascensus-acuminatus* Biozone, core BG-14, 36.0 m.

Fig. 4. *Normalograptus targuii* Legrand, 2001. BGS FOR 5399a, upper *ascensus-acuminatus* Biozone, core BG-14, 36.0 m.

Figs 5, 9. *Normalograptus parvulus* (H. Lapworth, 1900). 5, NHM Q6339(7), *persculptus* Biozone, Locality 5. 9, BGS FOR 5419b, upper *ascensus-acuminatus* Biozone, core BG-14, 39.8 m.

Fig. 6. *Neodiplograptus lanceolatus* Štorch and Serpagli, 1993. BGS FOR 5359d, upper *ascensus-acuminatus* Biozone, core BG-14, 43.4 m.

Fig. 7. *Paraclimacograptus libycus* (Desio, 1940). NHM QQ230, Aeronian, Locality 6, higher horizon (Aeronian).

Fig. 8. *Paraclimacograptus* sp. BGS FOR 5371a, *vesiculosus* Biozone, core BG-14, 30.0 m.

Fig. 10. *Metaclimacograptus hughesi* (Nicholson, 1969). BGS FOR 5363b, *vesiculosus* Biozone, core BG-14, 27.5 m.

All figures × 10.

PLATE 2

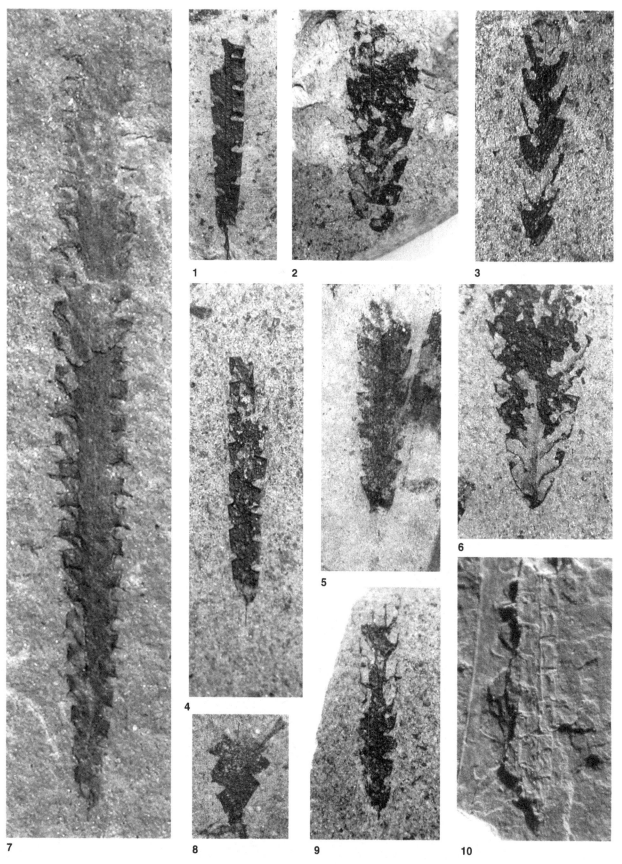

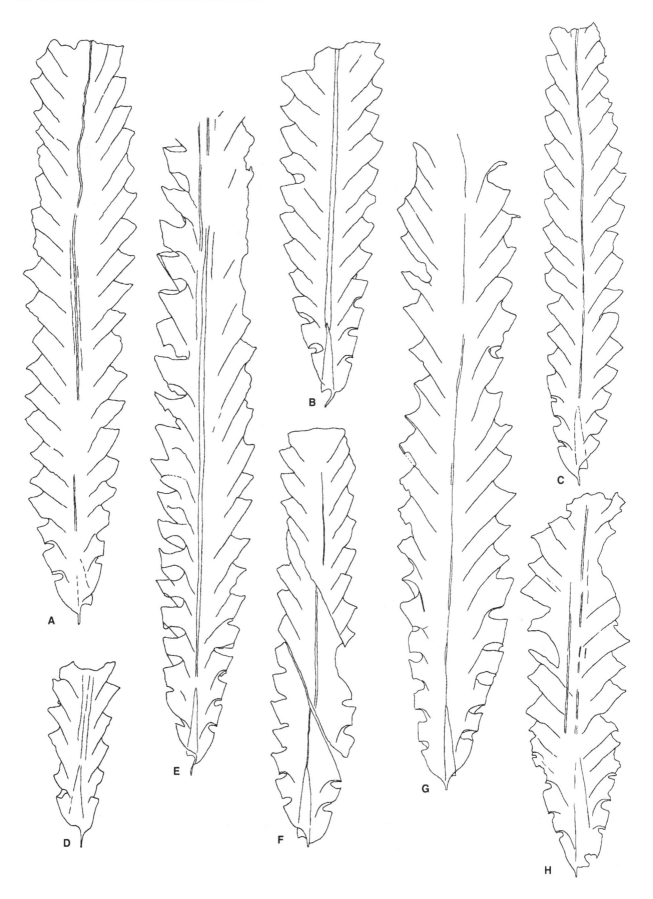

depths 9.1, 10.5, 10.77 and 13.8 m; one from Locality 1; and six from Locality 6 (lower horizon).

Description. As the name suggests, this species of *Neodiplograptus* often exhibits a lanceolate rhabdosome outline, being widest mesially. The sicula is 1.7–2.1 mm long ($n = 7$), its apex generally reaching to a level between the apertures of the second and third thecal pairs. The sicular aperture is 0.3–0.4 mm wide. The length of the exposed part of the dorsal sicular margin is 0.15–0.35 mm. Although generally short and inconspicuous, the virgella in a few specimens is longer, attaining a length of 2.5 mm, and is a little more robust than normal. Th1^1 grows downwards for 0.25–0.3 mm below the sicular aperture ($n = 3$). The ventral walls of the proximal thecal pair are inclined at *c.* 20 degrees to the rhabdosomal axis (measured just below the thecal aperture). Thecal apertures range from subhorizontal to everted. Measurements of DVW and 2TRD are given below. There is a complete median septum, although this is visible in only a few specimens.

DVW and 2TRD measurements (in mm)

th	1^1	2^1	3^1	5^1	10^1	15^1	20^1
DVW	0.95–1.4	1.2–1.8	1.35–2.1	1.95–2.7	2.5–3.35	2.5–3.4	3.0–3.3
2TRD		1.25–1.65	1.15–1.6	1.3–1.85	1.5–1.9	1.5–2.0	1.6–1.9

($n = 19$–21 for measurements at th1^1–5^1; $n = 15$ for measurements at th10^1; $n = 7$ for measurements at th15^1; $n = 3$ for measurements at th20^1).

Remarks. *Ne. lanceolatus* is very similar in appearance and DVW to '*Diplograptus*' *elacatus sensu* Fang *et al.* (1990). Note that the latter is significantly wider than Yu *et al.*'s (1988) original material of this species (see Chen *et al.* 2005a, p. 274). The specimens of Fang *et al.* (1990) may differ from *Ne. lanceolatus* in having more closely spaced thecae (2TRD is less than 1.5 mm throughout the length of the rhabdosome) and a more rounded proximal end; it is possible that the two are conspecific. With the exception of Tomczyk's (1962) Polish (Baltica) specimen and Maletz's (1999) specimen(s) from Belgium (Avalonia), and questionable occurrences in China (see synonymy above and that of Štorch and Serpagli 1993), all records of *Ne. lanceolatus* are from the margins of Gondwana.

One specimen (Text-fig. 11F), from a depth of 1399.5 m in core WS-6, has a proximal end identical to that of *Ne. lanceolatus* and increases in DVW over the first five thecal pairs at the same rate as this species. From the sixth thecal pair onwards, however, DVW decreases, to 1.9 mm at th10^1 and 1.8 mm at th15^1. This apparently

abnormal specimen is assigned questionably to *Ne. lanceolatus.*

Stratigraphical range. Štorch (1983a) found *Ne. lanceolatus* to be abundant in the lower and middle parts of the *ascensus-acuminatus* Biozone in Bohemia. A problem, however, with Bohemian sections is that the graptolite record is very sparse below the *ascensus* Biozone and thus it is not possible to assess whether those species, such as *Neodiplograptus lanceolatus*, that appear at the base of this biozone (Štorch 1994, fig. 2) in fact originated earlier, in the Late Ordovician, their absence from Bohemia simply reflecting local, unfavourable environmental or preservational conditions. In Jordan *Ne. lanceolatus* occurs with *Ne. apographon*, presumably in the upper part of the *ascensus-acuminatus* Biozone, although it does not occur in the highest 5.85 m of the biozone.

Neodiplograptus lueningi sp. nov.
Text-figures 12A, 13E–F

Derivation of name. After Sebastian Lüning.

Holotype. BGS FOR 5420a, core BG-14, depth 39.8 m (Text-figs 12A, 13E), from the upper *ascensus-acuminatus* Biozone.

Material. Three specimens, all from core BG-14, depths 39.8, 43.3 and 45.5 m.

Diagnosis. Robust *Neodiplograptus* attaining a DVW of 3.2 mm; proximal end is V-shaped; first eight thecal pairs are climacograptid; distal thecae are orthograptid.

Description. The rhabdosome is robust, attaining a DVW of 3.2 mm. The sicular apertural margin is damaged on all specimens. The sicula appears to have a length of 1.6–2.0 mm, the apex reaching the level either of the apertures of the second thecal pair or of th3^1. There is a robust virgella; this is broken 0.7 mm below the lowest point of th1^1. Measurements of DVW and 2TRD are given below. The proximal end is V-shaped. The first eight thecal pairs are of climacograptid appearance, with sharp genicula and supragenicular walls either subparallel to or inclined at less than 10 degrees to the rhabdosome axis. Thecal excavations are horizontal and deep, and semicircular to semioval. More distal thecae are of orthograptid appearance and are inclined at 35–40 degrees to the rhabdosome axis. They have everted apertures with slightly concave apertural margins. There is a complete median septum.

TEXT-FIG. 11. A–B, E, G, *Neodiplograptus lanceolatus* Štorch and Serpagli, 1993. A, BGS FOR 5465a, core WS-6, 1399.0 m, middle *ascensus-acuminatus* Biozone. B, BGS FOR 5464a, core WS-6, 1398.3 m, middle *ascensus-acuminatus* Biozone. E, BGS FOR 5427, core BG-14, 44.5 m, upper *ascensus-acuminatus* Biozone. G, NHM QQ235(1), Locality 6, lower horizon. C, *Neodiplograptus shanchongensis* (Li, 1984), BGS FOR 5466c, core WS-6, 1399.05 m, middle *ascensus-acuminatus* Biozone. D, *Sudburigraptus illustris* (Koren' and Mikhaylova, 1980), BGS FOR 5374a, core BG-14, 30.0 m, *vesiculosus* Biozone. F, *Neodiplograptus lanceolatus* Štorch and Serpagli, 1993?, BGS FOR 5471a, core WS-6, 1399.5 m, middle *ascensus-acuminatus* Biozone. H, *Neodiplograptus* sp. 2, BGS FOR 5369o, core BG-14, 30.0 m, *vesiculosus* Biozone. All figures × 10.

DVW and 2TRD measurements (in mm)

th	1^1	2^1	3^1	5^1	10^1	15^1	20^1
DVW	1.1–1.3	1.55–1.9	1.8–2.1	2.5	3.2	3.05	3.0
2TRD		1.3–1.55	1.45–1.6	–	1.7	1.6	

($n = 3$ up to th3^1, thereafter $n = 1$).

Remarks. Ne. lueningi has less inclined supragenicular walls in proximal thecae than *Ne. lanceolatus*. In this respect it resembles *Neodiplograptus mui* (Fang *et al.*, 1990) from the Hirnantian of China, but it differs in the orientation of the ventral walls of the first thecal pair, which are subparallel to the rhabdosome axis in *Ne. mui* but are inclined in *Ne. lueningi*. According to Fang *et al.*'s (1990, p. 122) description, *Ne. mui* is also broader, attaining a DVW of 3.5 mm at 5 mm from the proximal end.

Stratigraphical range. N. lueningi is known only from the upper *ascensus-acuminatus* Biozone of Jordan.

Neodiplograptus modestus (Lapworth, *in* Armstrong, Young and Robertson 1876)
Text-figures 12B, 13G

  1876 *Diplograptus modestus*, Lapw.; Lapworth, *in* Armstrong *et al.*, pl. 2, fig. 33.

  p 1907 *Diplograptus (Mesograptus) modestus*, Lapworth; Elles and Wood, p. 263, pl. 31, fig. 11*a–d* (*non* 11*e*); text-figs 180*a–b, d*) (*non* 180*c*).

  ? 1920 *Diplograptus modestus* Lapworth; Gortani, p. 21, pl. 1, fig. 34.

  ? 1940 *Diplograptus modestus* Lapworth; Desio, p. 25, pl. 1, fig. 18; pl. 2, figs 1, 5–6, 10, 12

  non 1962 *Diplograptus modestus modestus* Lapw.; Tomczyk, pl. 4, fig. 5; pl. 7, fig. 6.

  1970 *Diplograptus modestus modestus* Lapworth, 1876; Rickards, p. 35, text-fig. 13, fig. 4.

  1974 *Diplograptus modestus* Lapworth, 1876*a*; Hutt, p. 30, pl. 5, fig. 3; (?text-fig. 9, fig. 8).

  non 1983*a* *Diplograptus modestus modestus* Lapworth, 1876; Štorch, p. 162, pl. 1, figs 3, 5–6; text-fig. 2E–I [= *Neodiplograptus lanceolatus*].

  ? 1984 *Diplograptus modestus* Lapworth; Li, p. 342, pl. 5, fig. 9; pl. 12, figs 7–9; pl. 13, figs 1–4.

  non 1984 *Diplograptus modestus* Lapworth; Ge, p. 414, pl. 9, figs 17–18; text-fig. 4.

  non 1984 *Diplograptus modestus* Lapworth; Lin and Chen, pl. 2, figs 8–10; pl. 3, figs 11–12; pl. 5, fig. 7.

  non 1984 *Diplograptus modestus* Lapworth; Mu and Lin, p. 54, pl. 3, fig. 10.

  non 1986 *Diplograptus modestus* Lapworth; Štorch, pl. 5, fig. 2.

  non 1988 *Diplograptus modestus* Lapworth; Rickards, fig. 1m–n.

  1989 *Diplograptus modestus modestus* Lapworth; Melchin, fig. 6G.

  non 1990 *Diplograptus modestus* Lapworth; Gnoli *et al.*, text-fig. 4e.

  1994 '*Diplograptus*' *modestus modestus* Lapworth; Zalasiewicz and Tunnicliff, p. 712, text-fig. 9D–H.

  p 1998 *Neodiplograptus modestus modestus* (Lapworth); Underwood *et al.*, p. 104, fig. 5CC (*non* 5DD–EE).

  non 2003 *Neodiplograptus modestus modestus* (Lapworth); Masiak *et al.*, fig. 4m.

  non 2005*a* *Diplograptus modestus* Lapworth, 1876; Chen *et al.*, p. 273, text-fig. 11F.

Type specimen. Strachan (1997, p. 61) noted that no specimen matching Lapworth's original figure has been found. Material from Lapworth's original collection has been figured by Elles and Wood (1907) and Zalasiewicz and Tunnicliff (1994).

Material. One specimen, from the *vesiculosus* Biozone of Locality 7.

Description. The proximal end is poorly preserved. DVW increases steadily from *c.* 1.6 mm to 3.1 mm at the fifteenth thecal pair, after which this width initially is maintained; towards the distal end there is a slight decrease in DVW. 2TRD increases from 1.3–1.4 mm proximally to 1.7–1.85 mm distally. The thecae are geniculate throughout the length of the rhabdosome. The inclination of supragenicular walls to the rhabdosome axis declines from 25–40 degrees proximally to 10–28 degrees distally. The thecal apertures are everted proximally and horizontal distally. A median septum is present distally. Proximally, preservation is too poor to see whether one is present.

Remarks. Zalasiewicz and Tunnicliff (1994, text-fig. 9D–H) refigured specimens figured by Elles and Wood (1907) and additional material collected by Lapworth, all from the *atavus* (= lower *vesiculosus*) Biozone of Dob's Linn. Zalasiewicz and Tunnicliff (1994, p. 712) noted that the thecae have distinct genicula throughout the length of the rhabdosome. This character, together with the horizontal thecal apertures distally, distinguishes *Ne. modestus* from *Neodiplograptus lanceolatus* Štorch and Serpagli, 1993 with which it has frequently been confused (see Štorch and Serpagli 1993, p. 18 for discussion).

Chen *et al.* (2005*a*, p. 273) assigned two of Elles and Wood's (1907, pl. 31, fig. 11*d–e*) specimens to *Ne. charis* (Mu and Ni, 1983), a species from the Hirnantian *extraordinarius-osjuensis* Biozone of China. *Ne. charis* is, however, significantly broader (2.6–2.8 mm at the fifth to tenth thecal pair according to Chen *et al.* 2005*a*, p. 273) than these two specimens from the lower Llandovery, and has more widely spaced thecae.

Rickards (1988) illustrated two species under the name *Diplograptus modestus*. His figure 1m is *Cystograptus ancestralis* Štorch, 1985, matching precisely the dimensions and rhabdosome morphology of this taxon.

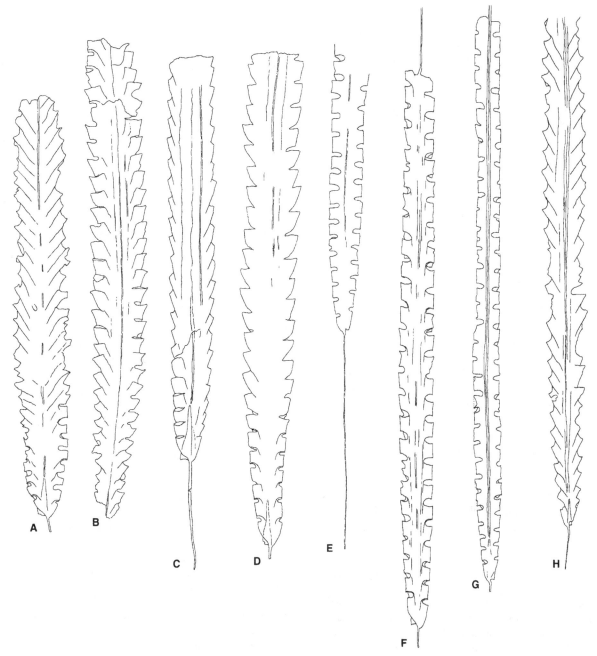

TEXT-FIG. 12.   A, *Neodiplograptus lueningi* sp. nov., holotype, BGS FOR 5420a, core BG-14, 39.8 m, upper *ascensus-acuminatus* Biozone. B, *Neodiplograptus modestus* (Lapworth, 1876), GD-NRA 18401-54d, Locality 7, *vesiculosus* Biozone. C–D, *Neodiplograptus parajanus* (Štorch, 1983a), core WS-6, middle *ascensus-acuminatus* Biozone. C, BGS FOR 5468a, 1399.3 m. D, BGS FOR 5472a, 1399.5 m. E–F, *Normalograptus normalis* (Lapworth, 1877). E, BGS FOR 5400a, core BG-14, 36.0 m, upper *ascensus-acuminatus* Biozone. F, holotype, BU 1136, *ascensus-acuminatus* Biozone of Dob's Linn, Scotland. G, *Normalograptus ajjeri* (Legrand, 1977), BGS FOR 5410c, core BG-14, 37.5 m, upper *ascensus-acuminatus* Biozone. H, *Sudburigraptus* sp., BGS FOR 5369q, core BG-14, 30.0 m, *vesiculosus* Biozone. All figures × 5.

The specimen illustrated by Chen *et al.* (2005a, text-fig. 11F) from the *persculptus* Biozone of Wangjiawan, China, differs from *Ne. modestus* in having a wider V-shaped proximal end and in the supragenicular walls of proximal thecae being inclined at a higher angle to the rhabdosome axis. It resembles closely *Ne. charis* (Mu and Ni, 1983), as illustrated by Chen *et al.* (2005a, text-fig. 10L).

Two of Underwood *et al.*'s (1998, fig. 5DD–EE) specimens have proximal ends that taper more than that of *Ne. modestus*. They are similar to specimens described from Algeria by Legrand (1995, p. 416, fig. 4) as *Ne.* aff. *africanus*.

Elles and Wood (1907, pl. 31, fig. 11a–d) illustrated considerable variation in the maximum DVW of *Ne. modestus*, from 2.2 to 3.0 mm. In this respect it is similar to *Ne. africanus* (Legrand, 1970). Elles and Wood (1907), Melchin (1989, fig. 6G) and Zalasiewicz and Tunnicliff (1994, text-fig. 9E) all illustrated a distally thickened nema, with this expanded into a nematularium in Melchin's specimen.

Stratigraphical range. Material from Scotland (Elles and Wood 1907), Arctic Canada (Melchin 1989) and Jordan is all from the lower part of the *vesiculosus* Biozone. Rickards (1970) and Hutt (1974) recorded *Ne. modestus* from this stratigraphical level, and also from the *acuminatus* Biozone *s.l.*

Neodiplograptus? opimus sp. nov.
Plate 1, figure 2; Text-figure 13D

? 1999 'Neodiplograptus' sp.; Maletz, fig. 2/10.

Derivation of name. Latin, *opimus*, fat.

Holotype. By monotypy, GD-NRA 8402-31d, a flattened rhabdosome, preserved in reverse view, from a depth of 46.0 m, BG-14 core, upper *ascensus-acuminatus* Biozone, Batna El Ghoul area, Jordan (Pl. 1, fig. 2; Text-fig. 13D).

Material. A single, flattened rhabdosome, from a depth of 46.0 m, BG-14 core.

Diagnosis. Neodiplograptus? with robust proximal end, 1.65 mm wide at $th1^1$. Rhabdosome appears almost parallel-sided. Thecae have subrounded to subangular genicula and horizontal apertures.

Description. The single rhabdosome is 7.3 mm long, excluding the virgella, and bears seven thecal pairs. It appears almost parallel sided. The sicular aperture is 0.5 mm wide and is furnished with a narrow virgella 1 mm long. The apex of the sicula cannot be clearly seen, but appears to reach approximately the level of the apertures of the second thecal pair. $Th1^1$ grows downwards for 0.35 mm below the sicular aperture. The length of the exposed part of the dorsal sicular margin is 0.35 mm. The thecae have subrounded to subangular genicula; distally this geniculum is somewhat less pronounced. Thecal apertures are horizontal. Proximal DVW and 2TRD measurements are given below. There is a complete median septum.

DVW and 2TRD measurements (in mm)

th	1^1	2^1	3^1	5^1	7^1
DVW	1.65	1.9	1.95	2.1	2.35
2TRD		1.6	1.6	1.9	

($n = 1$).

Remarks. Neodiplograptus? opimus is similar to *Neodiplograptus apographon* (Štorch, 1983a) in the rounded appearance of its proximal end. The latter species is much narrower proximally, however, and its distal changes in thecal morphology are more pronounced. Maletz's (1999, fig. 2/10) specimen matches *Ne.? opimus* in proximal DVW (1.6 mm at $th1^1$, 1.9 mm at $th2^1$), but is only questionably identified as the new species because it is only a very short, damaged specimen.

Stratigraphical range. Ne.? opimus is known only from the *ascensus-acuminatus* Biozone of Jordan.

Neodiplograptus parajanus (Štorch, 1983a)
Text-figures 12C–D, 14A–B, E

?p 1907   *Diplograptus (Mesograptus) modestus*, Lapworth; Elles and Wood, p. 263, pl. 31, fig. 11e (*non* 11a–d); text-fig. 180c (*non* 180a–b, d).

 ? 1965   *Diplograptus (Diplograptus?)* sp. B; Stein, p. 179, text-fig. 24b.

 *. 1983a   *Diplograptus parajanus* sp. n. Štorch, p. 168, pl. 4, figs 1–3; text-fig. 3A–B.

?p 1988   *Diplograptus modestus* Lapworth; Rickards, fig. 1n (*non* 1m).

   1993   *Neodiplograptus parajanus* (Štorch, 1983); Štorch and Serpagli, p. 19, pl. 3, figs 1, 5–6; text-figs 5A, F–G.

   1996   *Neodiplograptus parajanus* (Štorch); Štorch, fig. 4.4c.

Holotype. PŠ 54b, figured by Štorch (1983a, pl. 4, fig. 2, text-fig. 3A), from the Želkovice Formation of Praha-Řepy; *P. acuminatus* Biozone.

Material. 18 specimens: one from core BG-14, depth 42.6 m; 16 from core WS-6, depths 1399.3 and 1399.5 m; one from Locality 1.

Description. The length of the sicula is difficult to determine in the majority of specimens; in one it is 2.85 mm long, in another

TEXT-FIG. 13. A, *Neodiplograptus* sp. 5, BGS FOR 5464g, core WS-6, 1398.3 m, middle *ascensus-acuminatus* Biozone. B, H, *Neodiplograptus* sp. 1, core BG-14. B, BGS FOR 5437a, 46.3 m, upper *ascensus-acuminatus* Biozone. H, GD-NRA 8402-32a, 44.7 m, upper *ascensus-acuminatus* Biozone. C, I, *Neodiplograptus* sp. 4, Locality 5, *persculptus* Biozone. C, NHM Q 6339(10). I, Q6339(9). D, *Neodiplograptus? opimus* sp. nov., holotype, GD-NRA 8402-31d, core BG-14, 46.0 m, upper *ascensus-acuminatus* Biozone. E–F, *Neodiplograptus lueningi* sp. nov., core BG-14. E, holotype, BGS FOR 5420a, 39.8 m, upper *ascensus-acuminatus* Biozone. F, BGS FOR 5430d, 45.5 m, upper *ascensus-acuminatus* Biozone. G, *Neodiplograptus modestus* (Lapworth, 1876), GD-NRA 18401-54d, Locality 7, *vesiculosus* Biozone. All figures × 10.

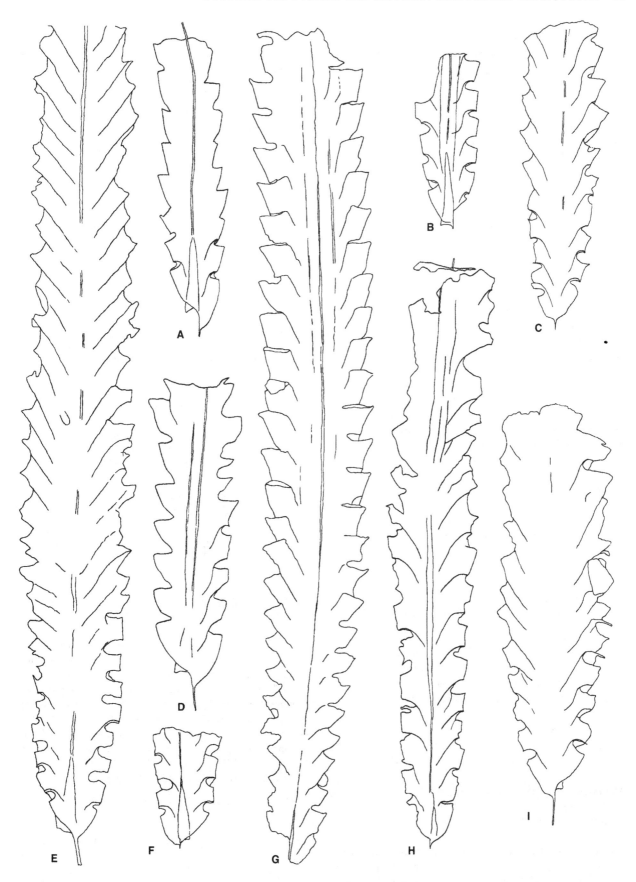

more than 2.3 mm long (it is damaged aperturally). The virgella is up to 7.5 mm long and usually robust. The appearance of the thecae varies significantly depending upon rhabdosome orientation. Proximal thecae have sharp genicula and supragenicular walls inclined at 2–15 degrees to the rhabdosome axis. At the fifth to seventh thecal pair thecal morphology changes to more glyptograptid or orthograptid. Thecal apertures are horizontal proximally; distally they are horizontal to slightly everted. Measurements of DVW and 2TRD are given below. There is a complete median septum.

DVW and 2TRD measurements (in mm)

th	1^1	2^1	3^1	5^1	10^1	15^1
DVW	1.2–1.5	1.3–1.65	1.5–1.8	1.8–2.2	2.5–2.6	2.5–2.6
2TRD		1.55–1.75	1.65–1.9	1.6–2.25	1.95–2.15	2.05–2.25

($n = 3$–5 for measurements at $th1^1$–10^1; $n = 2$ for measurements at $th15^1$ and $th20^1$).

Remarks. Some specimens (e.g. Text-figs 12C, 14E) of *Neodiplograptus parajanus* (Štorch, 1983a) appear similar to '*Orthograptus*'. *regularis* Fang et al., 1990 both in overall rhabdosome appearance and in the possession of a long robust virgella. The Chinese species differs, however, in having more closely spaced thecae (distal 2TRD is 1.6–1.7 mm) that are uniformly orthograptid and in possessing a broader nema.

Stratigraphical range. Štorch (1996, fig. 2; Text-fig. 6) indicated a range of middle–upper *acuminatus* Biozone for *Ne. parajanus*. The Jordanian occurrences are consistent with this; the WS-6 core specimens occur with *Ne. lanceolatus* and *A. ascensus* and are presumably from the lower part of the species' range.

Neodiplograptus shanchongensis (Li, 1984)
Text-figure 11C

> *. 1984 *Diplograptus shanchongensis* sp. nov. Li, p. 344, pl. 13, figs 9–11.
> ?p 1984 *Diplograptus modestus* Lapworth; Ge, p. 414, text-fig. 4 (*non* pl. 9, figs 17–18)
> p 1999 '*Neodiplograptus*' *lanceolatus* (Štorch and Serpagli, 1993); Maletz, p. 349, figs 3/1, 4/10 (?figs 2/9, 3/7, 4/7; *non* fig. 3/3).
> 2005a *Neodiplograptus shanchongensis* (Li, 1984a); Chen et al., p. 274, text-fig. 11b–e.

Lectotype. Not yet designated; Li's (1984) material is from the Kaochiapien Formation, Jiangxian, Anhui, China. This is stated to be of Hirnantian age by Chen et al. (2005a, p. 235).

Material. Two specimens: one bearing 17 thecal pairs, from a depth of 1399.05 m, core WS-6; the other bearing ten thecal pairs (eleventh pair damaged), from a depth of 16 m, core BG-4.

Description. The rhabdosome is 12 mm long in the longer specimen. The sicula has a length of at least 1.65 mm, its apex reaching above the level of the apertures of the second thecal pair.

The apertural width of the sicula is 0.25 mm. The dorsal margin of the sicula is exposed for 0.25 mm. $Th1^1$ grows downwards for 0.25 mm below the sicular aperture. The supragenicular walls of proximal thecae are straight or nearly so; distally the thecae become more glyptograptid to orthograptid in form, inclined to the rhabdosome axis at 35–40 degrees. Thecal apertures are everted throughout the length of the rhabdosome; thecal apertural margins are concave. Measurements of DVW and 2TRD are given below. There is a complete median septum.

DVW and 2TRD measurements (in mm)

th	1^1	2^1	3^1	5^1	10^1	15^1
DVW	1.0–1.1	1.2–1.45	1.45–1.7	1.8–2.0	2.25–2.3	1.8
2TRD		1.2–1.3	1.3–1.45	1.3–1.4	1.45	1.55

($n = 2$).

Remarks. *Ne. shanchongensis* differs from *Ne. lanceolatus* Štorch and Serpagli, 1993 in being narrower.

Stratigraphical range. Chen et al. (2005a, text-fig. 2) recorded *Ne. shanchongensis* as ranging throughout the *persculptus* Biozone and into the Rhuddanian of the Ludiping section. It occurs in the *acuminatus* Biozone of Belgium (Maletz 1999) and in Jordan in the *persculptus* Biozone in the BG-4 core and with *Akidograptus ascensus* in the WS-6 core.

Neodiplograptus sp. 1
Text-figures 13B, H

> v?p 1994 *Normalograptus? persculptus* (Elles and Wood, 1907); Zalasiewicz and Tunnicliff, p. 704, text-fig. 5A (?5C, *non* 5B).
> ?p 1998 *Persculptograptus persculptus s.l.* (Elles and Wood); Underwood et al., p. 103, fig. 5U (*non* 5S–T, V–W).

Material. Two proximal ends, from depths of 44.7 and 46.3 m, core BG-14.

Description. Both specimens have damaged proximal ends. In one the sicula is 2.0 mm long with an apertural width of 0.35 mm. The sicular apex reaches to the level of the top of $th2^2$. The proximal end is rounded. Proximally, the thecae have prominent horizontal apertural excavations and almost straight supragenicular walls, inclined at up to c. 10 degrees to the rhabdosome axis. Distal thecae are mostly damaged aperturally, but the apertures appear to be everted. Measurements of proximal DVW and 2TRD are given below. Maximum DVW is c. 2.2 mm. There is a complete median septum.

DVW and 2TRD measurements (in mm)

th	1^1	2^1	3^1	5^1
DVW	1.05–1.1	1.3–1.45	1.5–1.55	1.6
2TRD		1.3–1.55	1.3–1.4	1.45–1.5

($n = 2$).

Remarks. The appearance of the proximal end is unlike that of any other *Neodiplograptus* from the lower Llandovery. It is similar, both in dimensions and in morphology, to one of the proximal ends illustrated as '*Normalograptus? persculptus*' by Zalasiewicz and Tunnicliff (1994, text-fig. 5A). In this specimen, the thecae change from sharply geniculate proximally to more glyptograptid distally. One of Underwood *et al.*'s (1998, fig. 5U) specimens assigned to *N. persculptus* also appears very similar to the Jordanian material. *Ne.? imperfectus* (Legrand, 1986) differs from *Ne.* sp. 1 in having a narrower proximal end and more tapering rhabdosome proximally, and more widely spaced thecae.

Neodiplograptus sp. 2
Text-figure 11H

Material. One poorly preserved rhabdosome from a depth of 30.0 m, core BG-14.

Description. The rhabdosome is damaged proximally so that sicular details are not preserved. DVW increases from 1.35 mm at th1^1 to in excess of 2.5 mm at the seventh thecal pair. Proximal thecae are geniculate with straight supragenicular walls and conspicuous horizontal apertures. Distal thecae appear orthograptid, with everted apertures. They are inclined distally at 50 degrees to the rhabdosome axis. 2TRD increases from 1.3 mm at th2^1 to 1.5 mm distally. There appears to be a complete median septum.

Remarks. The specimen is similar in dimensions to *Ne. lanceolatus* Štorch and Serpagli, 1993, but differs in having its mesial thecae inclined at a higher angle (50 degrees compared with 30–40 degrees at the tenth thecal pair) to the rhabdosome axis. Comparison of the proximal ends is hampered by the poor preservation of this region in the single specimen of *Ne.* sp. 2.

Neodiplograptus sp. 3
Text-figures 14C–D, G

2000 *Nd. praeafricanus*; Legrand, fig. 3.

Material. Ten specimens: two from a depth of 27.5 m, one from 28.5 m and one from 30.0 m, core BG-14; one from a depth of 1399.5 m, core WS-6; two from a depth of 4.9 m, core BG-4; three from Locality 1.

Description. The rhabdosome is narrow and gently tapering. The sicular aperture is damaged in all but two specimens. Sicular length appears to be 1.2–1.8 mm, the apex reaching the level between the apertures of the first thecal pair to a little above the apertures of the second thecal pair. In the four specimens in which these features are visible, the dorsal margin of the sicula

is exposed for 0.15–0.3 mm and th1^1 grows downwards for 0.2–0.25 mm below the sicular aperture. The thecae have conspicuous apertural excavations and very gently inclined supragenicular walls (*c.* 10 degrees to the rhabdosome axis), having an almost climacograptid appearance. Thecal apertures are horizontal to slightly everted. Measurements of proximal DVW and 2TRD are given below. Two specimens decrease in DVW distally (e.g. Text-fig. 14G), one from 1.65 mm at th5^1 to 1.25 mm at th10^1 and 1.1 mm at th15^1. There appears to be a complete median septum.

DVW and 2TRD measurements (in mm)

th	1^1	2^1	3^1	5^1	10^1
DVW	0.75–0.9	0.9–1.15	1.05–1.25	1.25–1.65	1.4
2TRD		1.25–1.55	1.4–1.7	1.35–1.65	1.45

($n = 9$, except for th10^1 where $n = 1$).

Remarks. The material matches well that illustrated by Legrand (1999, 2000) in which distal thecae are shown to be glyptograptid. *Ne.* sp. 3 is narrower than *Ne. africanus* and has a more parallel-sided rhabdosome with thecae of more climacograptid appearance.

Stratigraphical range. Legrand (1999) considered this species stratigraphically to precede *Ne. africanus*. In Jordan it occurs with *Ne. africanus* in the BG-4 core and in the *vesiculosus* Biozone of the BG-14 core and also significantly lower in the Llandovery, with *A. ascensus*, in the WS-6 core.

Neodiplograptus sp. 4
Text-figures 13C, I

Material. Two specimens, both from Locality 5; *persculptus* Biozone.

Description. Both specimens are proximal ends, but neither preserves sicular details. The ventral walls of the first thecal pair are inclined at *c.* 30 degrees to the rhabdosome axis, giving the proximal end a distinctive V-shape. Proximal thecae are conspicuously geniculate; more distal thecae much less so, although in both specimens a geniculum is still apparent at the ninth thecal pair. Thecal apertures are everted. Distal thecae are inclined at 20–30 degrees to the rhabdosome axis. Dorso-ventral width increases from 1.25–1.6 mm at th2^1 to *c.* 3 mm distally. Both specimens have some damage to one series of thecal apertures, making precise DVW measurements difficult and giving the appearance of the presence of genicular hoods on some thecae. 2TRD increases from 1.45–1.5 mm at th2^2 to a maximum of 1.65 mm. A median septum is visible distally in one specimen; it is not clear whether it is complete.

Remarks. The proximal end is similar to that of *Ne. lueningi* sp. nov. The supragenicular walls are, however, inclined at a higher angle to the rhabdosome axis than in

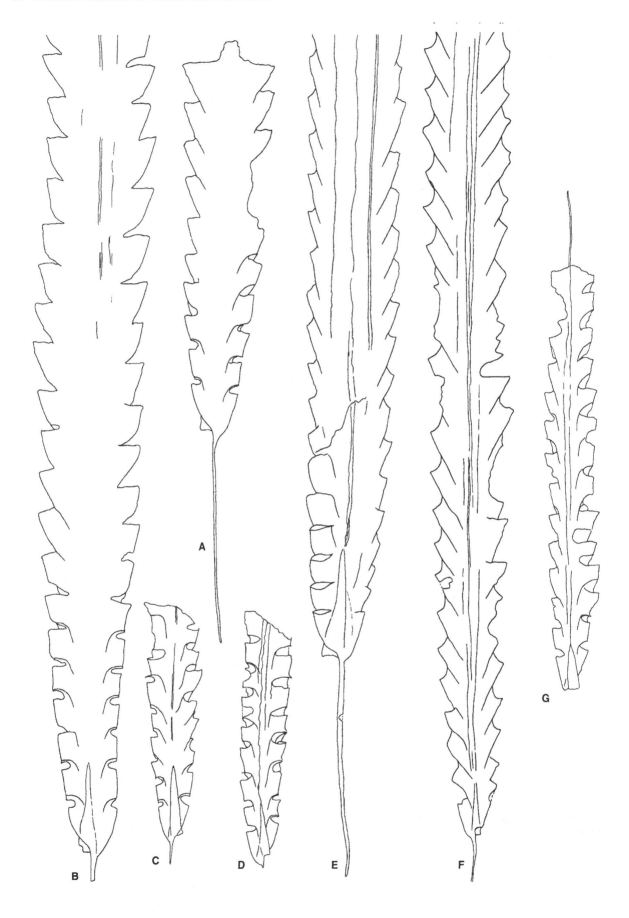

the latter species and the proximal thecal apertures are everted rather than being horizontal. The material is not well enough preserved for erection of a new species.

Neodiplograptus sp. 5
Text-figure 13A

Material. One proximal end from a depth of 1398.3 m, core WS-6.

Description. The sicula is 1.8 mm long, its apex reaching the aperture of th2^2. The aperture is 0.4 mm wide. The length of the exposed part of the dorsal sicular margin is 0.3 mm. Th1^1 grows down 0.5 mm below the sicular aperture. The rhabdosome increases in DVW proximally, but DVW decreases distally from th4^1. By comparison with an adjacent *Akidograptus ascensus*, the periderm appears to be thin. Genicula are subangular proximally and rounded distally. Supragenicular walls are inclined to the rhabdosome axis, at a higher angle proximally than distally. Measurements of DVW and 2TRD are given below. A median septum cannot be seen, but as most interthecal septa are not visible either, this should not be taken to indicate that *Ne.* sp. 5 is aseptate.

DVW and 2TRD measurements (in mm)

th	1^1	2^1	3^1	5^1	7^1
DVW	1.15	1.5	1.55	1.5	1.45
2TRD		1.55	1.45	–	–

($n = 1$).

Remarks. Neodiplograptus sp. 5 differs from other species in its combination of characters: distal decrease in DVW, long downward growth of th1^1 below the sicular aperture, and predominantly glyptograptid thecal morphology.

Genus NORMALOGRAPTUS Legrand, 1987, *emend.* Melchin and Mitchell, 1991

Normalograptus ajjeri (Legrand, 1977)
Text-figures 12G, 15A–F, H–I, O

vp 1906 *Climacograptus scalaris* (Hisinger) (Linné?) Var. *normalis*, Lapworth; Elles and Wood, p. 186, pl. 26, fig. 2*g* (?2*c, non* 2*a–b, d–f*); text-fig. 119*a* (?*b–d*).

1948 *Climacograptus scalaris* v. *normalis*; Wærn, p. 449, pl. 26, fig. 1; text-fig. 5.

1970 *Climacograptus normalis* Lapworth, 1877; Rickards, p. 28, pl. 1, fig. 1; text-fig. 13, figs 7–8.

1974 *Climacograptus normalis* Lapworth, 1877; Hutt, p. 19, pl. 1, figs 8–9; pl. 2, figs 1–2, 4.

*. 1977 *Climacograptus (Climacograptus) normalis ajjeri* nov. sub. sp. Legrand, p. 171, text-figs 9A–D, 10A–B.

1983 *Climacograptus normalis* Lapworth, 1877; Koren' *et al.*, p. 133, pl. 37, figs 1, 6–11; pl. 38, figs 1–5; (?pl. 39, fig. 7); text-fig. 48

p 1983 *Climacograptus normalis* Lapworth, 1877; Williams, p. 611, text-fig. 3*a–d*, 4*b–d*, 7*g* (?3*e*, 4*e*, *non* 4*a*).

1984 *Climacograptus normalis* Lapworth; Vandenberg *et al.*, fig. 10A–B.

1986 *Cl. normalis ajjeri*; Legrand, figs 2–3.

. 1986 *Climacograptus (Climacograptus) normalis brenansis* subsp. nov. Legrand, p. 151, fig. 5C–D.

. 1988 *Climacograptus normalis* Lapworth; Rickards, fig. 1b.

1988 *Climacograptus angustus* Perner; Cuerda *et al.*, fig. 6a–c.

p 1990 *Climacograptus medius* Tornquist; Fang *et al.*, p. 67, pl. 11, fig. 14 (*non* 5, 12–13).

1990 *Climacograptus normalis* Lapworth; Fang *et al.*, p. 69, pl. 12, fig. 4; pl. 13, figs 11–12.

1990 *Climacograptus scalaris* (Hisinger); Fang *et al.*, p. 70, pl. 12, figs 8–9.

. 1993 *Normalograptus normalis* (Lapworth); Štorch and Serpagli, p. 23, pl. 1, figs 3, 8; pl. 2, fig. 5; pl. 5, fig. 8; text-fig. 7L, N.

1994 *Normalograptus normalis* (Lapworth); Lenz and Vaughan, fig. 3A.

1996 *Normalograptus normalis* (Lapworth); Koren' and Rickards, pl. 5, figs 2–3, 6.

p 1996 *Normalograptus normalis* (Lapworth); Rickards *et al.*, p. 111, fig. 5F–G, 10C (*non* 10D–E).

. 1996 *Normalograptus normalis* (Lapworth); Štorch, fig. 4.3e.

p 2001 *Normalograptus (Normalograptus) normalis ajjeri* (Legrand, 1977); Legrand, p. 140, text-fig. 3a (?3b, *non* pl. 12, fig. 10).

. 2002 *N. (Normalograptus) normalis ajjeri* (Legrand); Legrand, pl. 12, figs 1–3; text-fig. 5a–c.

v. 2003 *Normalograptus normalis* (Lapworth); Loydell *et al.*, fig. 3a.

p 2003 *Normalograptus normalis* (Lapworth); Masiak *et al.*, figs 4e, 7b (?5c, *non* 5b).

. 2003 *N. normalis s.s.*; Koren' *et al.*, fig. 3:18–19.

TEXT-FIG. 14. A–B, E, *Neodiplograptus parajanus* (Štorch, 1983*a*), core WS-6, middle *ascensus-acuminatus* Biozone. A, BGS FOR 5468c, 1399.3 m. B, BGS FOR 5472a, 1399.5 m. E, BGS FOR 5468a, 1399.3 m. C–D, G, *Neodiplograptus* sp. 3. C, BGS FOR 5361b, core BG-14, 30.0 m, *vesiculosus* Biozone. D, BGS FOR 5471e, core WS-6, 1399.5 m, middle *ascensus-acuminatus* Biozone. G, BGS FOR 5457a, middle *ascensus-acuminatus* Biozone. F, *Sudburigraptus* sp., BGS FOR 5369q, core BG-14, 30.0 m, *vesiculosus* Biozone. All figures × 10.

Holotype. Original designation by Legrand (1977, p. 171): UL 2223 d9, from Oued In Djerane, Algeria.

Material. 42 specimens, from cores BG-4, BG-14 and WS-6, and Localities 4, 5, 7 and 13.

Description. The sicula is 1.5–2.15 mm long, with 14 of the 18 siculae measured being between 1.6 and 1.9 mm long. Apertural width is 0.3–0.35 mm. The length of the dorsal sicular margin exposed is very variable, between 0.15 and 0.5 mm. The sicular apex reaches from the aperture of $th2^1$ to just above that of $th2^2$. Virgellar length appears very variable, with a maximum length of 5.2 mm. $Th1^1$ grows downwards for 0.15–0.35 mm below the sicular aperture. The thecae are of typical climacograptid form, with sharp genicula and supragenicular walls parallel to the rhabdosome axis. The thecal apertures are horizontal. Measurements of DVW and 2TRD are given below. There is a complete median septum.

DVW and 2TRD measurements (in mm)

th	1^1	2^1	3^1	5^1	10^1	15^1
DVW	0.75–1.0	0.85–1.2	0.9–1.25	1.0–1.4	1.1–1.65	1.35–1.6
2TRD		1.2–2.15	1.3–2.1	1.35–1.95	1.5–1.95	1.65–1.9

($n = 29$–39 for $th1^1$–5^1; $n = 13$–17 for $th10^1$; $n = 4$–7 for $th15^1$).

Remarks. The description of Legrand's (1986) new species, *N. brenansis*, is almost identical to that of *N. ajjeri*; therefore, the species are considered synonymous. In *N. brenansis*, Legrand (1986, p. 152) had referred to the first theca's downward growth barely extending beyond the sicular aperture and the second theca ($th1^2$) crossing the sicula much higher above this aperture [than in *N. normalis*]. These features are shown by the holotype (Legrand 1986, fig. 5C). Legrand (1999, text-fig. 3-2-16a), however, figured a specimen (previously illustrated as Legrand 1986, fig. 5D; the 1999 figure is not a copy, but a new drawing) in which the downward growth of $th1^1$ extends significantly below the sicular aperture and in which $th1^2$ crosses the sicula only just above the aperture.

Stratigraphical range. As most previous records of *N. normalis* are *N. ajjeri*, this species clearly has a long stratigraphical range, from the Hirnantian through to the lower Aeronian.

Normalograptus angustus (Perner, 1895)
Plate 1, figure 3; Text-figure 15L–M

*. 1895 *Diplograptus (Glyptograptus) euglyphus* Lapworth, var. *angustus* mihi Perner, p. 27, pl. 8, fig. 14a–b.

?p 1906 *Climacograptus scalaris* (Hisinger) (Linné?) Var. *normalis*, Lapworth; Elles and Wood, p. 186, text-fig. 119*b* (*non* pl. 26, fig. 2*a–g*, text-fig. 119*a*, *c–d*).

vp 1906 *Climacograptus scalaris* (Hisinger) (Linné?) Var. *miserabilis* var. nov.; Elles and Wood, p. 186, pl. 26, fig. 3*d* (?3*a–b*, *e*, *non* 3*c*, *f–h*); text-fig. 120*a–b* (?120*c*).

. 1951 *Climacograptus angustus* (Perner, 1895); Přibyl, p. 7, pl. 2, figs 2–9.

p 1963 *Climacograptus angustus* (Perner 1895); Skoglund, p. 40, pl. 3, figs 1–2, 4–6; pl. 5, fig. 6 (*non* pl. 4, fig. 7).

1965 *Climacograptus scalaris miserabilis* Elles & Wood; Stein, p. 160, text-fig. 14g–h.

1970 *Climacograptus miserabilis* Elles and Wood, 1906; Rickards, p. 28, pl. 1, figs 3, 5, 10.

p 1970 *Climacograptus scalaris miserabilis* Elles & Wood; Toghill, p. 23, pl. 12, figs 1, 6, 9–11 (?5, *non* 2–4, 7–8).

. 1974 *Climacograptus miserabilis* Elles & Wood, 1906; Hutt, p. 20, pl. 1, figs 1–2; text-fig. 8, fig. 1.

. 1975 *Climacograptus angustus* (Perner, 1895); Bjerreskov, p. 23, fig. 9A.

. 1977 *Gl. (Climacograptus) miserabilis*, G. L. Elles et E. M. R. Wood; Legrand, text-fig. 7.

1977 *Climacograptus* sp. B; Legrand, p. 178, text-fig. 11A (?B).

non 1980 *Climacograptus angustus* (Perner, 1895); Koren' *et al.*, p. 131, pl. 37, figs 2–7; text-fig. 34 (all figures are *N. mirnyensis*).

1982 *Climacograptus miserabilis* Elles and Wood, 1906; Williams, p. 247, fig. 10k, n (?10l–m).

. 1983 *Climacograptus angustus* (Perner, 1895); Koren' *et al.*, p. 106, pl. 27, figs 1–5; text-fig. 34.

. 1983 *Climacograptus miserabilis* Elles and Wood, 1906; Williams, p. 615, text-figs 3*f–i*, 4*f–i*, 5*a–b*.

? 1984 *Climacograptus miserabilis* Elles and Wood; Li, p. 346, pl. 14, figs 7, 9, 12.

? 1984 *Climacograptus angustus* (Perner); Li, p. 350, pl. 15, figs 10–11.

TEXT-FIG. 15. A–F, H–I, O, *Normalograptus ajjeri* (Legrand, 1977), core BG-14. A, BGS FOR 5369k, 30.0 m, *vesiculosus* Biozone. B, BGS FOR 5429a, 44.5 m, upper *ascensus-acuminatus* Biozone. C, BGS FOR 5424, 43.4 m, upper *ascensus-acuminatus* Biozone. D, BGS FOR 5443c, 46.8 m, upper *ascensus-acuminatus* Biozone. E, BGS FOR 5417a, 39.0 m, upper *ascensus-acuminatus* Biozone. F, BGS FOR 5415, 37.5 m, upper *ascensus-acuminatus* Biozone. H, BGS FOR 5410a, 37.5 m, upper *ascensus-acuminatus* Biozone. I, BGS FOR 5409a, 37.5 m, upper *ascensus-acuminatus* Biozone. O, BGS FOR 5410c, 37.5 m, upper *ascensus-acuminatus* Biozone. G, J, *Normalograptus* cf. *normalis* (Lapworth, 1877), core BG-14. G, BGS FOR 5428a, 44.5 m, upper *ascensus-acuminatus* Biozone. J, BGS FOR 5423a, 43.4 m, upper *ascensus-acuminatus* Biozone. K, *Normalograptus trifilis* (Manck, 1923), BGS FOR 5401b, core BG-14, 36.8 m, upper *ascensus-acuminatus* Biozone. L–M, *Normalograptus angustus* (Perner, 1895), core BG-14. L, BGS FOR 5410d, 37.5 m, upper *ascensus-acuminatus* Biozone. M, BGS FOR 5438c, 46.8 m, upper *ascensus-acuminatus* Biozone. N, *Normalograptus cortoghianensis* (Štorch and Serpagli, 1993), BGS FOR 5400b, core BG-14, 36.0 m, upper *ascensus-acuminatus* Biozone. All figures × 10.

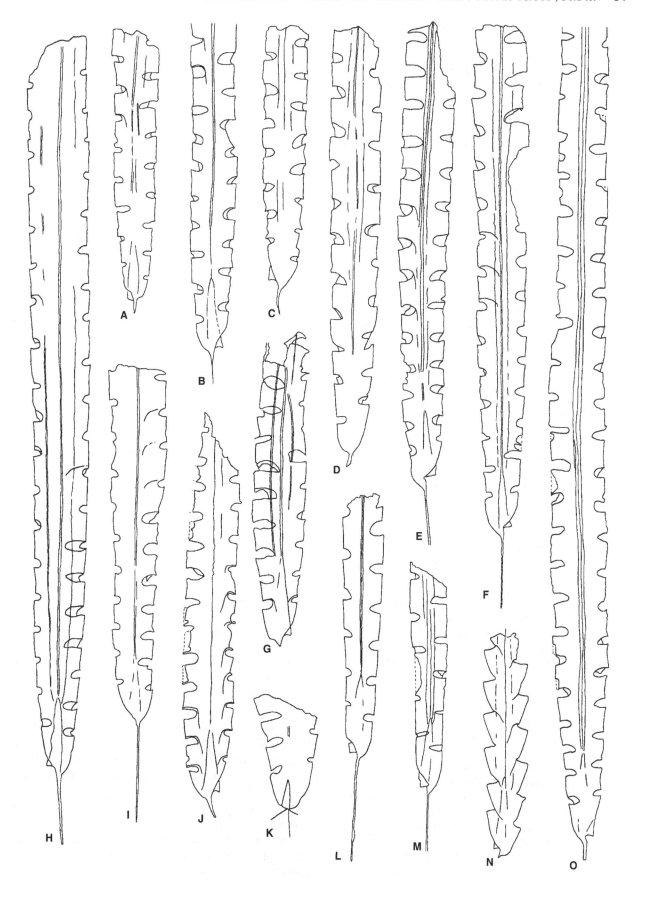

1984 *Climacograptus angustus* (Perner); Lin and Chen, pl. 6, fig. 3.

1984 *Climacograptus angustus* (Perner) (? = *C. miserabilis* Elles & Wood); Vandenberg *et al.*, fig. 4A.

1987 *Climacograptus miserabilis* Elles and Wood, 1906; Williams, p. 78, fig. 6A (?4G, 6B–C).

non 1988 *Climacograptus angustus* Perner; Cuerda *et al.*, fig. 6a–c.

. 1988 *Climacograptus angustus* Perner; Rickards, fig. 1i.

p 1988 *Scalarigraptus angustus* (Perner, 1895); Riva, p. 232, fig. 3a–b (?3c, non 3d–h, j–v) [see Chen *et al.* 2005a, p. 266].

non 1989 *Climacograptus angustus* (Perner); Melchin, fig. 5B (= *mirnyensis*).

. 1989 *Scalarigraptus angustus* (Perner, 1895); Štorch, p. 178, pl. 2, figs 3–5, 8; text-fig. 2E–J.

1990 *Climacograptus angustus* (Perner); Fang *et al.*, p. 65, pl. 14, figs 13–14 (?pl. 10, fig. 11).

. 1990 *Climacograptus miserabilis* (Elles et Wood); Fang *et al.*, p. 68, pl. 10, figs 2, 13 (?1).

p 1993 *Normalograptus angustus* (Perner, 1895); Štorch and Serpagli, pl. 1, figs 4, 6 (non 3 = *mirnyensis*); pl. 2, figs 2, 6; text-fig. 7A–B, E–F (non 7C = *mirnyensis*).

1994 *Normalograptus angustus* (Perner); Lenz and Vaughan, fig. 3B, D–E.

non 1994 *Normalograptus miserabilis* (Elles and Wood); Zalasiewicz and Tunnicliff, text-fig. 5κ–ʟ.

p 1994 *Normalograptus? parvulus* (H. Lapworth, 1900); Zalasiewicz and Tunnicliff, p. 705, text-fig. 5ɪ (non 5ᴅ–ʜ, ᴊ).

p 1996 *Normalograptus normalis* (Lapworth); Rickards *et al.*, p. 111, fig. 10D–E (non 5f–g, 10C).

. 1996 *Normalograptus angustus* (Perner, 1895); Štorch, fig. 4.3d.

1998 *Normalograptus ?mirnyensis* (Obut and Sobolevskaya); Underwood *et al.*, fig. 5P

. 1999 *Normalograptus angustus* (Perner, 1895); Maletz, p. 345, figs 3/5, 3/8, 4/5.

. 2000 *Normalograptus angustus* (Perner, 1895); Koren' and Melchin, p. 1097, figs 4.8, 4.11–4.13, 5.4.

?p 2000 *Normalograptus mirnyensis* (Obut and Sobolevskaya, 1967) new combination; Koren' and Melchin, p. 1099, fig. 7.13 (non 5.10, 5.17, 7.12, 8.1–8.5).

? 2001 *?Normalograptus* (*Normalograptus*) e.g. *miserabilis* (Elles et Wood, 1906); Legrand, p. 141, text-fig. 3c.

. 2003 *Normalograptus angustus* (Perner); Chen *et al.*, fig. 1Q.

2003 *Normalograptus miserabilis* (Elles and Wood); Masiak *et al.* fig. 4d (?4f, 7h).

2003 *Normalograptus mirnyensis* (Obut and Sobolevskaya); Chen *et al.*, fig. 1P (?1O).

. 2003 *Normalograptus angustus* (Perner); Chen *et al.*, fig. 1Q.

p 2003 *N. angustus*; Koren' *et al.*, figs 2.10–2.22, 3.13–3.14, 3.29 (non 3.11–3.12 = *mirnyensis*).

non 2003 *Normalograptus angustus* (Perner); Rong *et al.*, fig. 4.17 (= *mirnyensis*).

p 2005a *Normalograptus angustus* (Perner, 1895); Chen *et al.*, text-fig. 5D, DD (non 5I, K, Q = *mirnyensis*) [with additional synonymy].

Holotype. L27507, figured Perner (1895, pl. 8, fig. 14a–b), Přibyl (1951, pl. 2, fig. 8), Plate 1, figure 3, from the upper part of the Králův Dvůr Formation, Králův Dvůr, Bohemia.

Material. Ten specimens, from core BG-14.

Description. The sicula is 1.6–2.1 mm long ($n = 5$) with an apertural width of 0.25–0.3 mm. Its dorsal margin is exposed for 0.2–0.4 mm. Th1[1] grows downwards for 0.15–0.2 mm below the sicular aperture. The sicular apex reaches from just below the aperture of th2[1] to the aperture of th2[2]. The virgella has a maximum length of 3 mm. The thecae are of typical climacograptid form, with sharp genicula and supragenicular walls parallel to the rhabdosome axis or very gently inclined. The thecal apertures are horizontal. Measurements of DVW and 2TRD are given below. There is a complete median septum.

DVW and 2TRD measurements (in mm)

th	1[1]	2[1]	3[1]	5[1]
DVW	0.6–0.75	0.65–0.9	0.8–1.0	0.95–1.1
2TRD		1.8–2.2	1.75–2.5	2.15

($n = 5$ or 6, except for 2TRDth5[1] where $n = 2$). Note that the specimen with a 2TRD of 1.75 mm at th3[1] has a 2TRD at th2[1] of 2.2 mm and at th5[1] of 2.1 mm.

Remarks. Štorch (1989, p. 180) and Koren' and Melchin (2000, p. 1097) noted that the feature distinguishing *N. angustus* from *N. mirnyensis*, which possesses a similar DVW, is the former's more widely spaced thecae. On this basis, several specimens previously assigned to *N. angustus* are placed herein in *N. mirnyensis*.

Štorch and Serpagli (1993, p. 23) noted that Silurian material of *N. angustus* differed from that from the uppermost Ordovician in being narrower throughout is length. The dimensions of Chen *et al.*'s (2005a) Upper Ordovician specimens are, however, similar to those of younger material. Koren' and Melchin's (2000) specimens are conspicuously narrow proximally; of their five illustrated specimens only one exceeds 0.5 mm in DVW at th1[1]. Several authors, including Williams (1983, p. 615) and Štorch and Serpagli (1993, p. 23), have commented upon the variable appearance of the thecae, from climacograptid to glyptograptid.

Skoglund (1963, p. 40) described translucent isolated material of early growth stages of *N. angustus* and recorded a sicula 1.5 mm long. One of his figured specimens (pl. 4, fig. 7) shows closely spaced thecae and is therefore not included in *N. angustus*.

The specimen illustrated by Zalasiewicz and Tunnicliff (1994, text-fig. 5к) as *N. miserabilis* is very narrow distally (DVW 0.65 mm at th7[1]) and is similar in this respect to *N. skeliphrus* Koren' and Melchin, 2000, who quoted 2TRD measurements of 1.2–1.5 mm throughout the rhabdosome, rather lower than in Zalasiewicz and Tunnicliff's (1994) specimen. Koren' *et al.* (2003, fig. 3.19), however, assigned a specimen with more widely spaced thecae to *N. skeliphrus* and on this basis Zalasiewicz and Tunnicliff's (1994) specimen may also be accommodated within this species.

Riva (1988) studied the type material of *N. angustus* and *N. miserabilis* and concluded that they were synonyms. Although this view is now generally accepted, examination of Elles and Wood's (1906) figured material indicates that possibly three taxa were illustrated by them as *miserabilis*. Strachan (1971, p. 81) recognized that some specimens differed from typical *miserabilis* and assigned them to 'C. cf. *supernus* Elles and Wood'. Of the remainder, BU 1145b (Elles and Wood 1906, pl. 26, fig. 3b) possesses a tapering rhabdosome and attains a DVW of 1.45 mm, much broader than recorded by Williams (1987, p. 80) even in tectonically distorted specimens from the same locality and horizon as Elles and Wood's material. Some of the other specimens have more closely spaced thecae than is typical for *angustus*. They may be tectonically deformed or possibly *N. mirnyensis*. However, as Williams (1987, p. 79) selected BU 1150 (Elles and Wood 1906, text-fig. 120a; refigured Riva 1988, fig. 3b) as lectotype of *C. miserabilis*, its synonymy with *angustus* still stands.

Stratigraphical range. *Normalograptus angustus* has a very long stratigraphical range, from the *Dicranograptus clingani* Biozone of the Upper Ordovician (Williams 1982; as *Climacograptus miserabilis*) through to the Rhuddanian: both Rickards (1970) and Hutt (1974) recorded it from the *cyphus* Biozone.

Normalograptus bifurcatus sp. nov.
Plate 1, figure 5; Text-figure 16A, D–E, J

Holotype. NHM Q6337(3) (Pl. 1, fig. 5; Text-fig. 16D), from low hill *c.* 2 km west of Jebel Badra, near the main highway between Ma'an and Mudawwara; Mudawwara Formation; *persculptus* Biozone.

Material. 16 specimens with proximal ends, plus numerous other fragments, from Locality 5 and core BG-4, depth 16 m.

Diagnosis. *Normalograptus* with sicula bearing both a bifurcating virgella and a bifurcating antivirgellar spine.

Description. The sicula is 1.5–2.05 mm long ($n = 2$) with an apertural width of 0.25–0.3 mm. Its dorsal margin is exposed for 0.15–0.5 mm ($n = 5$). Th1[1] grows downwards for 0.05–0.35 mm

below the sicular aperture. The sicular apex reaches the aperture of th2[1]. The virgella is long and straight. It bifurcates at a distance from the sicular aperture varying between 0.15 mm (i.e. at the base of the downward growing portion of th1[1]) and 1.5 mm. The angle between the two virgellar branches is 5–15 degrees. There is an antivirgellar spine that also bifurcates, with an angle of 5–15 degrees between the branches, at a variable distance from the sicular aperture, between 0.5 and 2.6 mm. The thecae are of typical climacograptid form, with sharp genicula and supragenicular walls parallel to the rhabdosome axis or very gently inclined. The thecal apertures are horizontal. Measurements of DVW and 2TRD are given below; the relatively small number of measurements reflects the high proportion of scalariform and subscalariform views. There appears to be a complete median septum.

DVW and 2TRD measurements (in mm)

th	1[1]	2[1]	3[1]	5[1]	10[1]
DVW	0.65–0.75	0.7–0.85	0.7–0.9	0.9–1.0	1.0–1.25
2TRD		1.55–1.85	1.65–1.8	1.6–2.0	1.75–2.25

($n = 5$ or 6 up to th3[1]; $n = 3$ or 4 at th5[1]; $n = 2$ or 3 at th10[1]).

Remarks. This new species resembles *N. minor* (Huang, 1982) (see Chen *et al.* 2005a, pl. 2, figs 1–2; pl. 3, figs 1, 3–4; text-fig. 8D–E, J, O, R, T) in having a bifurcating virgella. It differs in the narrower angle between the two virgellar branches, in the straightness of the branches and in the possession of a long bifurcating spine originating from the dorsal margin of the sicular aperture.

Stratigraphical range. At Locality 5, *N. bifurcatus* occurs with *N. persculptus*, indicating the *persculptus* Biozone. Its occurrence at 16 m in core BG-4 (Text-fig. 2) is also consistent with this biozone as this is below strata yielding the *ascensus-acuminatus* Biozone indicator *Neodiplograptus lanceolatus*.

Normalograptus cortoghianensis (Štorch and Serpagli, 1993)
Plate 2, figure 3; Text-figure 15N

*. 1993 *Glyptograptus cortoghianensis* n. sp. Štorch and Serpagli, p. 32, pl. 2, fig. 1; pl. 4, fig. 4; text-fig. 5B, D.
1996 *Glyptograptus cortoghianensis* Štorch and Serpagli; Štorch, fig. 4.1a.

Holotype. By original designation, UMPI 22586a, figured Štorch and Serpagli (1993, pl. 2, fig. 1; text-fig. 5D) from the *P. acuminatus* Biozone of Monte Cortoghiana Becciu, Sardinia.

Material. One specimen from a depth of 36.0 m in core BG-14.

Description. The single rhabdosome is 6 mm long. The sicula is 1.8 mm long, its apex reaching the aperture of th2[2]. The sicula's dorsal margin is exposed for 0.2 mm. Th1[1] grows downwards

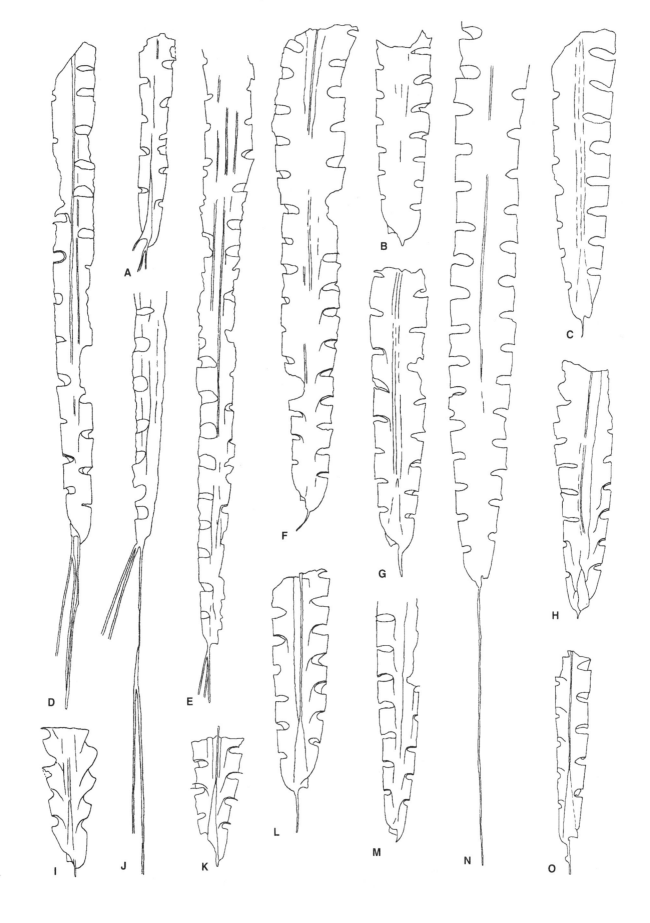

for 0.2 mm below the sicular aperture. The proximal end appears rounded and the thecae are glyptograptid throughout the length of the rhabdosome. Thecal apertures are slightly everted. Measurements of DVW and 2TRD are given below. There is a complete median septum.

DVW and 2TRD measurements (in mm)

th	1^1	2^1	3^1	5^1
DVW	0.9	0.95	1.15	1.2
2TRD		1.65	1.75	1.75

($n = 1$).

Remarks. Štorch and Serpagli (1993) originally assigned *N. cortoghianensis* tentatively to *Glyptograptus* Lapworth, 1873. *Glyptograptus* is aseptate, however (Melchin 1998, p. 298), whilst both Sardinian and Jordanian material of *N. cortoghianensis* possesses a median septum. Štorch and Serpagli (1993, p. 5) found *N. cortoghianensis* to be a rare component of early Silurian assemblages from Sardinia. It is also rare in Jordan.

Stratigraphical range. In Sardinia, *N. cortoghianensis* is known from throughout the *P. acuminatus* Biozone. The Jordanian specimen is from the upper part of this biozone.

Normalograptus larini Koren' and Melchin, 2000
Plate 2, figure 1; Text-figure 16O

? 1996 *Normalograptus* cf. *balticus* (Pedersen, 1922); Koren' and Rickards, p. 37, pl. 4, fig. 10.
*. 2000 *Normalograptus larini* new species Koren' and Melchin, p. 1099, figs 5.7, 7.1–7.6

Holotype. GSC118372 (Koren' and Melchin 2000, fig. 7.1), from sample III-1/5, Urubulak Formation, Kurama Range, Uzbekistan; *Hirsutograptus sinitzini* Subzone, *Akidograptus ascensus-Parakidograptus acuminatus* Biozone.

Material. One specimen from a depth of 36.8 m, core BG-14; upper *ascensus-acuminatus* Biozone.

Description. The rhabdosome is preserved in obverse view. It tapers conspicuously towards the proximal end. The sicula is

1.8 mm long. Its dorsal margin is exposed for 0.2 mm. Th1[1] grows downwards for 0.2 mm below the sicular aperture. The sicular apex attains a level just above the top of th2[1]. The virgella is robust; it is broken after a length of 0.8 mm. The thecae are typically climacograptid, with sharp genicula and supragenicular walls parallel to the rhabdosome axis or very gently inclined. The thecal apertures are horizontal. Measurements of DVW and 2TRD are given below. There appears to be a complete median septum.

DVW and 2TRD measurements (in mm)

th	1^1	2^1	3^1	5^1
DVW	0.65	0.75	0.85	1.0
2TRD		1.6	1.6	1.65

($n = 1$).

Remarks. The specimen agrees well with Koren' and Melchin's (2000) description. *Normalograptus larini* differs from *N. mirnyensis* in its more conspicuously tapering proximal end and particularly in its possession of a robust virgella.

Stratigraphical range. Koren' and Melchin (2000, table 1) recorded *N. larini* only from the *sinitzini* Subzone of the *ascensus-acuminatus* Biozone. The Jordanian specimen described herein is from the upper *ascensus-acuminatus* Biozone, a rather higher biostratigraphical level (see discussion of Koren' and Melchin's *P. acuminatus* below).

Normalograptus medius (Törnquist, 1897)
Text-figures 16B, F–G, L, 17A–B

v*. 1897 *Climacograptus medius* n. sp. Törnquist, p. 7, pl. 1, figs 9–15.
non 1940 *Climacograptus medius* Törnquist; Desio, p. 27, pl. 1, figs 16–17.
non 1948 *Cl. medius*; Wærn, p. 450, pl. 26, fig. 4; text-fig. 5.
? 1965 *Climacograptus medius* Törnquist, 1897; Stein, p. 163, text-fig. 16a–g.
1970 *Climacograptus medius* Törnquist, 1897; Rickards, p. 30, pl. 1, fig. 2.
. 1974 *Climacograptus medius* Törnquist, 1897; Hutt, p. 19, pl. 1, fig. 3.

TEXT-FIG. 16. A, D–E, J, *Normalograptus bifurcatus* sp. nov., Locality 5, *persculptus* Biozone. A, NHM Q6338(5). D, holotype, NHM Q6337(3). E, NHM Q6341(1). J, NHM Q6337(2). B, F–G, L, *Normalograptus medius* (Törnquist, 1897). B, BGS FOR 5443a, core BG-14, 46.8 m, upper *ascensus-acuminatus* Biozone. F, BGS FOR 5468f, core WS-6, 1399.3 m, middle *ascensus-acuminatus* Biozone. G, BGS FOR 5370v, core BG-14, 30.0 m, *vesiculosus* Biozone. L, BGS FOR 5421b, 42.5 m, upper *ascensus-acuminatus* Biozone. C, H, N, *Normalograptus normalis* (Lapworth, 1877), core BG-14. C, BGS FOR 5360g, 36.0 m, upper *ascensus-acuminatus* Biozone. H, BGS FOR 5378a, 32.5 m, upper *ascensus-acuminatus* or *vesiculosus* Biozone. N, BGS FOR 5400a, 36.0 m, upper *ascensus-acuminatus* Biozone. I, *Normalograptus* sp., BGS FOR 5369 m, core BG-14, 30.0 m, *vesiculosus* Biozone. K, *Paraclimacograptus obesus* (Churkin and Carter, 1970), BGS FOR 5375d, core BG-14, 30.0 m, *vesiculosus* Biozone. M, *Normalograptus transgrediens* (Wærn, 1948), BGS FOR 5368a, core BG-14, 28.5 m, *vesiculosus* Biozone. O, *Normalograptus larini* Koren' and Melchin, 2000, BGS FOR 5407b, core BG-14, 36.8 m, upper *ascensus-acuminatus* Biozone. All figures × 10.

. 1975 *Climacograptus medius* Törnquist, 1897; Bjerreskov, p. 24, fig. 9C.

1977 *Cl. (Climacograptus) medius*, S. L. Törnquist; Legrand, text-fig. 1D (?1C).

1979 *Climacograptus* cf. *medius* Törnquist; Jaeger and Robardet, pl. 2, fig. 19.

. 1983 *Climacograptus medius* Törnquist, 1897; Williams, p. 616, text-fig. 5f–i.

? 1988 *Climacograptus medius* Törnquist; Cuerda *et al.*, fig. 5k–l.

. 1988 *Climacograptus medius* Törnquist; Rickards, fig. 1a.

non 1990 *Climacograptus medius* Tornquist; Fang *et al.*, p. 67, pl. 11, figs 5, 12–14.

? 1993 *Normalograptus medius* (Törnquist, 1897); Štorch and Serpagli, p. 23, pl. 5, figs 1, 7; text-fig. 7D, O.

1995 *Normalograptus medius* (Törnquist); Piçarra *et al.*, fig. 3-5.

1996 *Normalograptus medius* (Tornquist); Štorch, fig. 4.3g.

non 1998 *Normalograptus medius*; Underwood *et al.*, fig. 5J.

1999 *Normalograptus medius* (Törnquist, 1897); Maletz, p. 340, fig. 3/6.

p 2003 *Normalograptus normalis* (Lapworth); Masiak *et al.*, fig. 5b (*non* 4e, 5c, 7b).

2003 *Normalograptus medius* (Törnquist); Masiak *et al.*, fig. 5d (?e).

non 2005a *Normalograptus medius* (Törnquist, 1897); Chen *et al.*, p. 260, pl. 2, fig. 5; text-fig. 8C.

Lectotype. Designated by Přibyl (1948, p. 16), the specimen figured by Törnquist (1897, pl. 1, fig. 9), from the 'Rastrites Beds' of Nyhamn, Scania, Sweden; *cyphus* Biozone. The lectotype is figured as Text-figure 17A.

Material. 14 specimens from Locality 7 and cores BG-4, BG-14 and WS-6.

Description. The proximal end is conspicuously rounded, resulting from the convexity of the ventral walls of the first thecal pair. The sicula is 1.6–1.8 mm long ($n = 3$). Apertural width is 0.3–0.4 mm. The length of the exposed dorsal sicular margin is 0.1–0.4 mm ($n = 9$). The sicular apex reaches to the aperture of th2^2. The virgella is up to 9 mm long. Th1^1 grows downwards for 0.1–0.3 mm below the sicular aperture ($n = 11$; in nine the distance is 0.2–0.25 mm). The thecae are of typical climacograptid form, with sharp genicula and supragenicular walls parallel to the rhabdosome axis. The thecal apertures are deep and horizontal. Measurements of DVW and 2TRD are given below. There is a complete median septum.

DVW and 2TRD measurements (in mm)

th	1^1	2^1	3^1	5^1	10^1	15^1
DVW	0.9–1.1	1.15–1.2	1.1–1.4	1.3–1.75	1.6–1.85	1.8–1.95
2TRD		1.4–1.75	1.45–1.65	1.45–1.75	1.45–2.0	1.85–2.0

($n = 7$–9 for th1^1–5^1; $n = 5$ for th10^1; and $n = 3$ or 4 for th15^1).

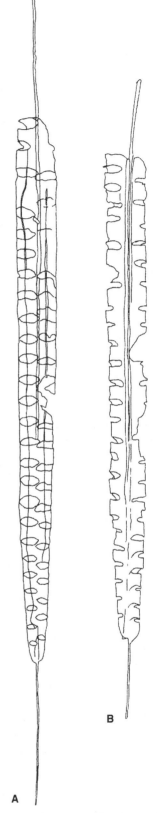

B

A

TEXT-FIG. 17. *Normalograptus medius* (Törnquist, 1897), *cyphus* Biozone, Nyhamn, Scania, Sweden. A, lectotype, LO 1262T. B, specimen on same slab as lectotype. Both × 5.

Remarks. Normalograptus medius has a characteristically rounded proximal end and it is this character that distinguishes it from *N. normalis*. Bjerreskov (1975, p. 24) compared her material from Bornholm with Törnquist's original specimens and described the proximal end as 'stout and rounded'. Legrand (1977, fig. 1D), however, illustrated a 'fac similé du lectotype' showing a tapering proximal end merging with a stout virgella in the manner of *N. balticus* (Pedersen, 1922). The illustration in Legrand (1977) is rather misleading, however. The lectotype is illustrated herein as Text-figure 17A and shows a long, narrow virgella with no evidence for this merging with the proximal end in the manner shown in Legrand (1977). Přibyl's (1948) choice of lectotype was unfortunate, as Törnquist (1897, explanation to pl. 1, fig. 9) stated that this is a 'compressed scalariform specimen' whereas the specimen illustrated in his plate 1, figure 10 is a biprofile view. The latter, however, has a damaged proximal end (pers. obs.). A specimen from the slab bearing the lectotype is shown in Text-figure 17B and shows the typically rounded proximal end.

Chen *et al.*'s (2005*a*) material is not included in *N. medius* as the figured specimens have inclined supragenicular walls, shallow thecal excavations and rather blunt, rather than rounded, proximal ends.

Stein (1965) distinguished *N. medius* from *N. rectangularis* (McCoy, 1850) by the presence in the former of a long virgella. Virgellar length is highly variable, however, both in *N. medius* and in other *Normalograptus* species. The tapering rhabdosome of *N. rectangularis* is the most useful distinguishing feature. Törnquist (1897, pl. 1, fig. 12) illustrated a specimen with a 12-mm-long virgella.

Stratigraphical range. Normalograptus medius is a long-ranging species, recorded from the upper *extraordinarius* Biozone (of north-east Russia; Rickards 1988, p. 345) through to the lower Aeronian *triangulatus* Biozone (Rickards 1970).

Normalograptus mirnyensis (Obut and Sobolevskaya, in Obut *et al.* 1967)
Text-figure 18A–E

vp?	1906	*Climacograptus scalaris* (Hisinger) (Linné?) Var. *miserabilis* var. nov.; Elles and Wood, p. 186, pl. 26, fig. 3*a–b, e* (*non* 3*c–d, f–h*); text-fig. 120*a–b* (?120*c*).
*.	1967	*Hedrograptus mirnyensis* Obut et Sobolevskaya, sp. nov.; Obut *et al.*, p. 47, pl. 1, figs 4–9.
p	1970	*Climacograptus scalaris miserabilis* Elles & Wood; Toghill, p. 23, pl. 12, figs 2, 4, 7–8 (?5, *non* 1, 3, 6, 9–11).
?	1977	*Climacograptus* (*Climacograptus*) sp. C; Legrand, p. 180, text-fig. 12A–C.

	1980	*Climacograptus angustus* (Perner, 1895); Koren' *et al.*, p. 131, pl. 37, figs 2–7; text-fig. 34.
.	1980	*Climacograptus mirnyensis* (Obut et Sobolevskaya, 1967); Koren' *et al.*, p. 137, pl. 39, figs 1, 3; text-fig. 39.
	1983	*Climacograptus mirnyensis* (Obut et Sobolevskaya, 1967); Koren' *et al.*, p. 132, pl. 37, figs 2–5; text-fig. 47k–n.
	1984	*Climacograptus mirnyensis* (Obut et Sobolevskaya); Ge, p. 427, pl. 8, fig. 5.
?	1984	*Climacograptus miserabilis* Elles et Wood: Ge, p. 428, pl. 8, figs 1–2; pl. 9, fig. 4.
	1984	*Climacograptus mirnyensis* (Obut et Sobolevskaya); Li, p. 347, pl. 14, figs 8, 11.
?p	1987	*Climacograptus miserabilis* Elles and Wood, 1906; Williams, p. 78, fig. 6B (?4G, 6C, *non* 6A).
?p	1988	*Scalarigraptus angustus* (Perner, 1895); Riva, p. 232, fig. 3c (*non* 3a–b, d–h, j–v) [see Chen *et al.* 2005a, p. 266 for identifications].
	1989	*Climacograptus minutus* Carruthers; Melchin, fig. 5A.
	1989	*Climacograptus angustus* (Perner); Melchin, fig. 5B.
?	1990	*Climacograptus mirnyensis* (Obut et Sobolevskaya); Fang *et al.*, p. 68, pl. 13, figs 8–10.
p	1993	*Normalograptus angustus* (Perner, 1895); Štorch and Serpagli, pl. 1, fig. 3 (*non* pl. 1, figs 4, 6; pl. 2, figs 2, 6); text-fig. 7C (*non* 7A–B, E–F).
	1998	*Normalograptus miserabilis* (Elles and Wood); Underwood *et al.*, fig. 5F (?5E).
.	2000	*Normalograptus mirnyensis* (Obut and Sobolevskaya, 1967) new combination; Koren' and Melchin, p. 1099, figs 5.10, 5.17, 7.12 (?7.13), 8.1–8.5.
p	2003	*N. angustus*; Koren' *et al.*, fig. 3.11–3.12 (*non* 2.10–2.22, 3.13–3.14, 3.29).
	2003	*Normalograptus mirnyensis* (Obut et Sobolevskaya); Rong *et al.*, fig. 4.3, 8, 14, 19 (?6).
	2003	*Normalograptus angustus* (Perner); Rong *et al.*, fig. 4.17.
p	2005a	*Normalograptus angustus* (Perner, 1895); Chen *et al.*, text-fig. 5I, K, Q (*non* 5D, DD).
.	2005a	*Normalograptus mirnyensis* (Obut et Sobolevskaya, in Obut *et al.* 1967); Chen *et al.*, p. 262, text-fig. 8B, F–G, I, L–N [with further synonymy].

Holotype. By original designation, figured by Obut *et al.* (1967, pl. 1, fig. 4), from the *ascensus-acuminatus* Biozone of Mirny Creek, north-east Russia.

Material. 13 specimens, from core BG-14 and Localities 7 and 13.

Description. The sicula is 1.3–1.8 mm long (*n* = 10) with an apertural width of 0.25–0.35 mm. Its dorsal margin is exposed for 0.1–0.4 mm. $Th1^1$ grows downwards for 0.1–0.25 mm below the sicular aperture. The sicular apex reaches to just below the aperture of $th2^1$ up to the aperture of $th2^2$. The virgella has a maximum length of 6.5 mm. The thecae are of typical climacograptid form, with sharp genicula and supragenicular walls parallel to the rhabdosome axis or very gently inclined. The thecal apertures are horizontal. Measurements of DVW and 2TRD are given below. There is a complete median septum.

DVW and 2TRD measurements (in mm)

th	1^1	2^1	3^1	5^1	10^1
DVW	0.65–0.8	0.7–0.85	0.7–1.0	0.75–1.05	0.95–1.1
2TRD		1.45–1.7	1.5–1.75	1.45–1.9	1.6–1.65

($n = 9$–12 up to $th5^1$; $n = 2$ or 3 at $th10^1$).

Remarks. Some collections of *N. mirnyensis* contain specimens with short virgellae, up to 1 mm long (e.g. Koren' and Melchin 2000; Chen *et al.* 2005a). Others contain much longer virgellae (e.g. Koren' *et al.* 1983), approaching 3 mm long. The Jordanian collection includes specimens with much longer virgellae, but in all other respects the material is identical to previously described specimens.

Stratigraphical range. Chen *et al.* (2005a, p. 262) recorded *N. mirnyensis* throughout the Hirnantian; it also occurs in the lower Rhuddanian.

Normalograptus normalis (Lapworth, 1877)
Plate 1, figures 4, 7; Text-figures 12E–F, 16C, H, N, 19

v*. 1877 *Climacograptus scalaris.* His. (Non Linnæus.) Var. *b. normalis.* Lapw., Lapworth, p. 138, pl. 6, fig. 31.

vp 1906 *Climacograptus scalaris* (Hisinger) (Linné?) Var. *normalis*, Lapworth; Elles and Wood, p. 186, pl. 26, fig. 2a, d, f (? 2c, non 2b, e, g); ?text-fig. 119c (*non* 119a–b, d).

non 1948 *Climacograptus scalaris* v. *normalis*; Wærn, p. 449, pl. 26, fig. 1; text-fig. 5.

? 1965 *Climacograptus scalaris normalis* Lapworth, 1877; Stein, p. 157, pl. 14, fig. C; text-figs 13i, 14a–e.

non 1970 *Climacograptus normalis* Lapworth, 1877; Rickards, p. 28, pl. 1, fig. 1; text-fig. 13, figs 7–8.

non 1974 *Climacograptus normalis* Lapworth, 1877; Hutt, p. 19, pl. 1, figs 8–9; pl. 2, figs 1–2, 4.

non 1976 *Hedrograptus normalis* (Lapworth, 1877); Sennikov, p. 133, pl. 4, figs 12–13.

non 1978 *Climacograptus normalis* Lapworth; Baillie *et al.*, p. 47, fig. 3.

non 1979 *Climacograptus normalis* Lapworth, 1877; Paškevičius, p. 108, pl. 1, figs 1–4; pl. 18, figs 4–5.

1982 *Climacograptus normalis* Lapworth; Lenz and McCracken, fig. 4c–d.

non 1983 *Climacograptus normalis* Lapworth; Mu and Ni, p. 168, pl. 2, fig. 5.

non 1983 *Climacograptus normalis* Lapworth, 1877; Koren' *et al.*, p. 133, pl. 37, figs 1, 6–11; pl. 38, figs 1–5; pl. 39, fig. 7; text-fig. 48.

p 1983 *Climacograptus normalis* Lapworth, 1877; Williams, p. 611, text-fig. 4a (?3e, *non* 3a–d, 4b–e, 7g).

non 1984 *Climacograptus normalis* Lapworth; Lin and Chen, pl. 3, fig. 10.

non 1984 *Climacograptus normalis* Lapworth; Vandenberg *et al.*, fig. 10A–B.

. 1988 *Climacograptus normalis* Lapworth; Cuerda *et al.*, fig. 6d.

non 1988 *Climacograptus normalis* Lapworth; Rickards, fig. 1b.

non 1988 *Scalarigraptus normalis* (Lapworth); Riva, fig. 3.

non 1989 *Climacograptus normalis* Lapworth; Melchin, fig. 7E.

p 1990 *Climacograptus medius* Tornquist; Fang *et al.*, p. 67, pl. 11, figs 5, 13 (*non* 12, 14).

1991 *Normalograptus normalis*; Gutiérrez-Marco and Robardet, fig. 2h.

non 1993 *Normalograptus normalis* (Lapworth); Štorch and Serpagli, p. 23, pl. 1, figs 3, 8; pl. 2, fig. 5; pl. 5, fig. 8; text-fig. 7L, N.

non 1994 *Normalograptus normalis* (Lapworth); Lenz and Vaughan, fig. 3A.

non 1994 *Normalograptus normalis* (Lapworth); Zalasiewicz and Tunnicliff, text-fig. 6o.

non 1996 *Climacograptus normalis* Lapworth; Rickards *et al.*, p. 111, figs 5f–g, 10C–E.

TEXT-FIG. 18. A–E, *Normalograptus mirnyensis* (Obut and Sobolevskaya, *in* Obut *et al.*, 1967), core BG-14, upper *ascensus-acuminatus* Biozone. A, BGS FOR 5409b, 37.5 m. B, BGS FOR 5421a, 42.5 m. C, BGS FOR 5420i, 39.8 m. D, BGS FOR 5416a, 39.0 m. E, BGS FOR 5417c, 39.0 m. F–G, K, P–Q, S, *Normalograptus targuii* Legrand, 2002, core BG-14, upper *ascensus-acuminatus* Biozone. F, BGS FOR 5404, 36.8 m. G, BGS FOR 5406a, 36.8 m. K, BGS FOR 5401a, 36.8 m. P, BGS FOR 5399g, 36.0 m. Q, BGS FOR 5360h, 36.0 m. S, BGS FOR 5399a, 36.0 m. H, R, T, *Normalograptus rectangularis* (McCoy, 1850). H, BGS FOR 5378b, core BG-14, 32.5 m, upper *ascensus-acuminatus* or *vesiculosus* Biozone. R, BGS FOR 5461, Locality 10, *vesiculosus* Biozone. T, BGS FOR 5380b, core BG-14, 32.5 m, upper *ascensus-acuminatus* or *vesiculosus* Biozone. I–J, L–M, O, *Normalograptus parvulus* (H. Lapworth, 1900). I, BGS FOR 5447a, Locality 1, upper *ascensus-acuminatus* Biozone. J, BGS FOR 5465h, core WS-6, 1399.0 m, middle *ascensus-acuminatus* Biozone. L, BGS FOR 5446, Locality 1, upper *ascensus-acuminatus* Biozone. M, NHM QQ235(3), Locality 6, lower horizon. O, NHM Q6339(7), Locality 5, *persculptus* Biozone. N, *Normalograptus persculptus* (Elles and Wood, 1907), NHM 6339(6), Locality 5, *persculptus* Biozone. All figures × 10.

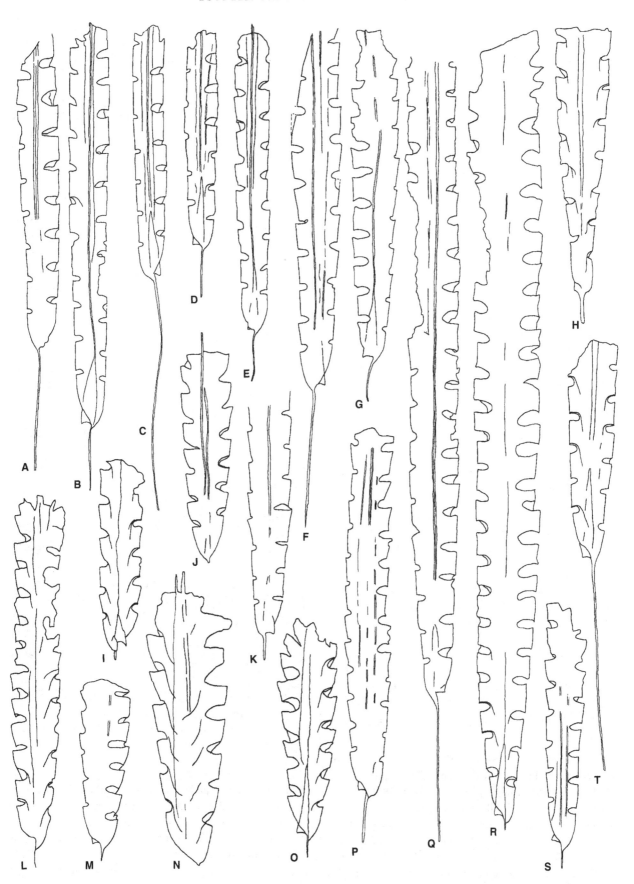

non 1996   *Normalograptus normalis* (Lapworth); Koren'
     and Rickards, pl. 5, figs 2–3, 6.
non 1996   *Normalograptus normalis* (Lapworth); Štorch,
     fig. 4.3e.
non 1998   *Normalograptus normalis* (Lapworth, 1877);
     Melchin, text-fig. 3c.
non 1999   *Normalograptus normalis* (Lapworth, 1877);
     Maletz, p. 344, fig. 2/15 (= *N. rectangularis*).
non 2000   *Normalograptus normalis* (Lapworth, 1877);
     Koren' and Melchin, p. 1101, fig. 7.14.
  . 2003   *N. normalis* (a wider form), Koren' *et al.*, fig.
     3.3–4, 25–27.
non 2003   *Normalograptus normalis* (Lapworth); Loydell
     et al., fig. 3a.
non 2003   *Normalograptus normalis* (Lapworth); Masiak
     et al., figs 4e, 5b–c, 7b.
non 2003   *Normalograptus normalis* (Lapworth); Rong
     et al., fig. 4.7.
non 2005a *Normalograptus normalis* (Lapworth, 1877);
     Chen *et al.*, p. 264, text-fig. 9J.

Holotype. BU 1136, figured by Elles and Wood (1906, pl. 26, fig. 2*a*), Williams (1983, text-fig. 4*a*), Text-figures 12F, 19; from the *acuminatus s.l.* Biozone of Dob's Linn, Moffat, southern Scotland.

Material. 11 specimens, from cores BG-4, BG-14, and Localities 7 and 13. One specimen from a depth of 32.5 m in core BG-14 is preserved in low relief.

Description. The sicula is 1.5–1.85 mm long ($n = 5$). Apertural width is 0.25–0.35 mm. The length of the dorsal sicular margin exposed is 0.25–0.35 mm ($n = 8$; in seven specimens it is 0.25 mm). The sicular apex reaches to the apertures of the second thecal pair. The virgella is up to 11 mm long. Th1^1 grows downwards for 0.15–0.25 mm below the sicular aperture ($n = 8$; in five the distance is 0.25 mm). The thecae are of typical climacograptid form, with sharp genicula and supragenicular walls parallel to the rhabdosome axis. The thecal apertures are horizontal. Measurements of DVW and 2TRD are given below. There is a complete median septum.

DVW and 2TRD measurements (in mm)

th	1^1	2^1	3^1	5^1	10^1	15^1
DVW	0.8–1.05	0.95–1.2	1.1–1.4	1.3–1.75	1.6–1.85	1.7–1.95
2TRD		1.4–1.75	1.4–1.65	1.45–1.75	1.45–2.0	1.85–2.1

($n = 9$–11 for th1^1–5^1; $n = 7$ for th10^1; and $n = 4$ or 5 for th15^1).

Remarks. Lapworth's (1877, p. 138) original description was very brief: 'Polypary with subparallel margin. Virgula greatly prolonged distally.' Elles and Wood (1906, p. 186) added a little more detail, but characterized *N. normalis* as 'never exceeding 1.5 mm in breadth'. This characteristic has been oft repeated in diagnoses for the species (e.g. Rickards 1970, p. 28; Hutt 1974, p. 19; Williams 1983, p. 613). However, the holotype and several of Elles and

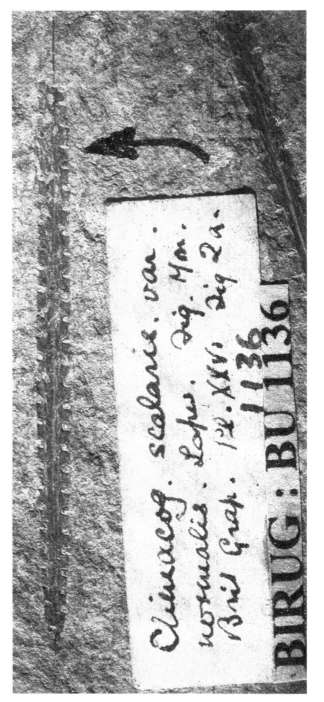

TEXT-FIG. 19. *Normalograptus normalis* (Lapworth, 1877), holotype, BU 1136, *ascensus-acuminatus* Biozone, Dob's Linn, Scotland; × 5.

Wood's figured specimens attain a DVW significantly greater than 1.5 mm: the holotype, for example, has a maximum DVW of 1.85 mm. A specimen (BU 1136e) on the same slab as the holotype attains a maximum DVW of 2.05 mm. Williams (1983, p. 613) considered the holotype to be 'tectonically widened'; the orientation of its thecal apertures suggests some deformation, but dimensions are unlikely to have been altered dramatically (Mike Melchin, pers. comm. December 2006). This is suggested also by the fact that BU 1136b, which is at 50 degrees to the holotype, also attains a DVW of 1.85 mm.

Koren' et al. (2003) recognized what they called 'a wider form' of N. normalis from the uppermost Ordovician and lowermost Silurian of Scania, Sweden. The DVW of this material matches that of the holotype and the broader material illustrated by Elles and Wood (1906) and should therefore be assigned to N. normalis s.s.

Two of Elles and Wood's specimens (Elles and Wood 1906, pl. 26, fig. 2b: specimen stated by Strachan 1971, p. 81 to be SM A20057, but examination of the latter indicates that this is not this figured specimen; and BU 1140: Elles and Wood 1906, pl. 26, fig. 2g; see below for measurements of DVW and 2TRD) are Normalograptus premedius (Wærn, 1948), a species he recorded (p. 456) from Sweden from the base of the acuminatus Biozone and by Koren' et al. (2003) from the upper persculptus Biozone, also of Sweden. This species is characterized by a tapering rhabdosome with widely spaced thecae and a median septum commencing on the reverse side of the rhabdosome at the level of the third to fourth thecal pairs (that on BU 1140 commences at the level of the aperture of th3[1]). Zalasiewicz and Tunnicliff's (1994) specimen may also be N. premedius; the specimen has a tapering rhabdosome and the thecae are very widely spaced (2TRD is 2.5 mm at th5[1]).

DVW and 2TRD measurements of N. premedius, BU 1140 (in mm)

th	1[1]	2[1]	3[1]	5[1]	10[1]	15[1]	20[1]
DVW	0.6	0.65	0.75	0.95	1.25	1.45	1.35
2TRD		1.95	2.1	2.4	2.5	2.75	2.55

Chen et al. (2005a, p. 264) commented on the wide, blunt proximal end of their Chinese material and for this reason it is excluded from N. normalis herein. This is significant as Mitchell et al. (2003, p. 525) and Chen et al. (2005b, p. 856) noted that in all sequences except those of the Yangtze Platform, N. normalis became extinct during the early Hirnantian, reappearing as a Lazarus taxon 'with the return of graptolite-rich black shales in the late Hirnantian or early Rhuddanian'. In the Yangtze Platform sections N. normalis was stated to be present throughout the Hirnantian. This clearly is not the case. Underwood et al. (1998, fig. 5C) illustrated a similar, blunt proximal-ended specimen from Mauritania, indicating that this form (which may be a new species) may have been geographically widespread.

Strachan (1971, p. 101) stated that the specimen illustrated by Elles and Wood (1906) as text-fig. 119a is BU 1141. This specimen is, however, the counterpart to the holotype, BU 1136, and is much broader than indicated by Elles and Wood's text-figure.

Paškevičius (1979) illustrated material from the Aeronian of Lithuania. The thecal apertures are much narrower than those of N. normalis. These specimens can be assigned to N. scalaris (Hisinger, 1837) (see Rickards 1973, fig. 5 for illustration of lectotype).

Normalograptus cf. normalis (Lapworth, 1877)
Text-figure 15G, J

cf. 1877 Climacograptus scalaris. His. (Non Linnæus.) Var. b. normalis. Lapw., Lapworth, p. 138, pl. 6, fig. 31.
  1993 Normalograptus normalis (Lapworth)?; Štorch and Serpagli, pl. 1, fig. 7; text-fig. 7M.
p 1998 Normalograptus ex gr. normalis (Lapworth); Underwood et al. p. 103, fig. 5A–B (non 5C).
  1998 Normalograptus sp.? normalis (Lapworth); Underwood et al., p. 103, fig. 5D.

Material. 13 specimens, from core BG-14.

Description. The length of the sicula can be measured in two specimens only, in which it is 1.65 mm and 2.0 mm. The sicular apertural width is 0.3–0.35 mm. The length of the exposed part of the dorsal sicular margin is 0.25–0.5 mm ($n = 8$). Th1[1] grows downwards for 0.2–0.35 mm below the sicular aperture ($n = 7$). DVW increases up to th5[1] at a rate similar to that in N. normalis. From this point on, however, DVW decreases distally. Measurements of DVW and 2TRD are given below. There is a complete median septum.

DVW and 2TRD measurements (in mm)

th	1[1]	2[1]	3[1]	5[1]	10[1]
DVW	0.85–1.2	1.0–1.4	1.15–1.7	1.3–1.75	1.25–1.6
2TRD		1.3–1.55	1.4–1.6	1.4–1.75	1.45–1.9

($n = 11$–13 for measurements to th5[1]; $n = 9$ for DVWth10[1]; $n = 6$ for 2TRDth10[1]).

Remarks. Similar fusiform Normalograptus specimens have been recorded from Sardinia (Štorch and Serpagli 1993) and Mauritania (Underwood et al. 1998).

Normalograptus parvulus (H. Lapworth, 1900)
Plate 2, figures 5, 9; Text-figure 18I–J, L–M, O

*. 1900 Climacograptus parvulus H. Lapworth, p. 132, fig. 20a–c.
p 1907 Mesograptus modestus Var. parvulus (H. Lapworth); Elles and Wood, p. 264, pl. 31, fig. 12a (non 12b–d); text-fig. 181a (non 181b–c).

. 1929   *Glyptograptus persculptus* mut. omega. nov.; Davies, p. 14, figs 15, 20.

non 1940   *Diplograptus modestus* Lapworth var. *parvulus* (Lapworth); Desio, p. 27, pl. 2, fig. 4.

p 1974   *Glyptograptus (G.) persculptus* (Salter, 1865); Hutt, p. 28, pl. 6, figs 9–12.

? 1977   *Diplograptus modestus parvulus* (H. Lapworth); Legrand, p. 188, text-fig. 16.

1983a   *Diplograptus* aff. *parvulus* (H. Lapworth); Štorch, p. 164, pl. 4, fig. 4; text-fig. 2O.

. 1983   *Glyptograptus* cf. *persculptus* (Salter, 1865); Williams, p. 625, pl. 66, figs 4–7.

non 1984   *Diplograptus parvulus* (H. Lapworth); Li, p. 343, pl. 13, fig. 12.

1987   *Glyptograptus persculptus* Salter; Wang, pl. 1, fig. 1.

. 1988   *Glyptograptus persculptus* (Salter), *sensu lato*; Rickards, fig. 1o–p.

?p 1988   *Scalarigraptus angustus* (Perner, 1895); Riva, p. 232, fig. 3d–h, n–s (*non* 3a–c, j–m, t–v).

vp 1994   *Normalograptus? persculptus* (Elles and Wood, 1907); Zalasiewicz and Tunnicliff, p. 704, text-fig. 5B (?5C, *non* 5A).

p 1994   *Normalograptus? parvulus* (H. Lapworth, 1900); Zalasiewicz and Tunnicliff, p. 705, text-fig. 5D–H, J (*non* 5I).

1996   *Glyptograptus persculptus* (Salter); Rickards *et al.*, p. 110, figs 5a–b, d, (?c), 9a–b.

1996   ?*Climacograptus ?ojsuensis*; Rickards *et al.*, fig. 5e.

vp 1996   *Normalograptus persculptus* (Elles and Wood); Štorch and Loydell, p. 872, text-fig. 3E (*non* 3A–D, F–H, 4–5).

p 1998   *Persculptograptus persculptus s.l.* (Elles and Wood); Underwood *et al.*, p. 103, fig. 5V (*non* 5S–U, W).

v. 2002   *Normalograptus parvulus* (Lapworth, 1900); Loydell *et al.*, p. 136, figs 2–3; table 1.

? 2003   *Normalograptus parvulus* (H. Lapworth); Masiak *et al.*, figs 4g, l, q, 7e–f.

. 2005a   *Normalograptus parvulus* (Lapworth, 1900); Chen *et al.*, p. 266, text-fig. 9E, H, O, S.

v. 2005   *Normalograptus parvulus* (Lapworth, 1900); Armstrong *et al.*, fig. 6a–d.

Lectotype. BU 1293, figured H. Lapworth, 1900, fig. 20*b*; refigured Elles and Wood 1907, pl. 31, fig. 12*a* and text-fig. 181*a* and by Zalasiewicz and Tunnicliff 1994, text-fig. 5D; from the lower Cwmere Formation, Gwastaden, Rhayader, central Wales. Elles and Wood (1907, explanation to pl. 31) and Strachan (1997, p. 61) stated that the specimen is from the *acuminatus* Biozone. Zalasiewicz and Tunnicliff (1994, p. 705), however, stated that it is from either the *persculptus* or the *acuminatus* Biozone.

Material. More than 100 specimens; preserved flattened with the exception of the specimens from Localities 4 and 9 and those from Locality 7 figured by Armstrong *et al.* (2005), which are preserved in partial to full relief. The species is present at all localities except Localities 2 and 10.

Description. The sicula is 1.35–1.8 mm long ($n = 17$), with an apertural width of 0.25–0.4 mm ($n = 18$). The length of the exposed part of the dorsal sicular margin is 0.15–0.45 mm (24 of the 30 measurements made were between 0.2 and 0.35 mm). The virgella is short and inconspicuous. Th1[1] grows downwards for 0.1–0.4 mm below the sicular aperture (29 of the 32 measurements made were between 0.15 and 0.3 mm). Measurements of DVW and 2TRD are given below. A few specimens (e.g. Text-fig. 18L) show a distal decrease in DVW. The appearance of the thecae is somewhat variable, depending upon orientation and preservation style. Genicula, although generally sharp, occasionally appear more rounded, presenting a more glyptographid profile to the thecae. Supragenicular walls are inclined to the rhabdosome axis, this character being least pronounced in relief material (e.g. Armstrong *et al.* 2005, fig. 6; Text-fig. 18I). Details of the median septum are not visible in all specimens; in some it appears complete.

DVW and 2TRD measurements (in mm)

th	1[1]	2[1]	3[1]	5[1]	10[1]	15[1]	20[1]
DVW	0.7–1.0	0.8–1.2	0.9–1.35	1.05–1.55	1.25–1.7	1.4–1.6	1.5
2TRD		1.15–1.65	1.2–1.8	1.25–1.8	1.45–1.7	1.4–1.75	1.6–1.65

($n = 42$–62 for measurements to th5[1]; $n = 13$ for th10[1]; $n = 5$ for th15[1]; and $n = 2$ for th20[1]).

Remarks. The ranges of measurements of proximal DVW and 2TRD in *Normalograptus parvulus* and *N. persculptus* overlap and, as a result, Zalasiewicz and Tunnicliff (1994, p. 705) suggested that *N. parvulus* may be an end-member of a single, variable species that also includes *N. persculptus*. However, distally the two species are clearly separable, with *N. persculptus* having significantly higher DVW and/or 2TRD values (see Text-fig. 9).

Davies (1929) recognized that his material from the *acuminatus* Biozone of Wales differed from *N. persculptus* from the *persculptus* Biozone, but suggested that the younger forms (his mut. omega) 'may be found to be synonymous with *Mesograptus* [now *Neodiplograptus*] *modestus*'. He did not mention H. Lapworth's species *parvulus*. As shown by Zalasiewicz and Tunnicliff (1994, text-fig. 9D–H), *Ne. modestus* has a very different thecal morphology and widens much more rapidly than *N. parvulus*. Rickards (1988) refigured Davies's (1929) specimens; they are clearly *N. parvulus*.

One of the specimens illustrated by Zalasiewicz and Tunnicliff (1994, text-fig. 5B) as *Normalograptus persculptus* is assigned here to *N. parvulus*: the thecae are very closely spaced proximally (2TRD at th2[1] is 1.2 mm; at th5[1] is 1.4 mm). In *N. persculptus* the thecae are more widely spaced (e.g. 1.5–1.8 mm at th2[1] in the specimens studied by Štorch and Loydell 1996; 1.65–1.95 mm at th2[1] in the specimens illustrated by Koren' *et al.* 2003). In Loydell *et al.*'s (2002) material of *N. parvulus* 2TRD at th2[1] was 1.15–1.6 mm.

Similarly, Hutt's (1974, pl. 6, figs 9–12) specimens identified as '*G. persculptus*' are narrow and have closely spaced thecae and can be placed in *N. parvulus*. It is possible that *N. persculptus* was present in Hutt's collection as she refers (1974, p. 28) to a single specimen with a width of 1.9 mm.

The specimen illustrated by Zalasiewicz and Tunnicliff (1994, text-fig. 5ɪ) as *Normalograptus parvulus* is *N. angustus* (Perner, 1895). The supragenicular walls are parallel (or very nearly so) to the rhabdosome axis, the rhabdosome does not increase sufficiently rapidly in dorso-ventral width and the thecae are too widely spaced (2TRD at th2^1 is 1.9 mm).

Zalasiewicz and Tunnicliff (1994, p. 705) referred to 'moderately well-developed genicular hoods'. These are not visible in their figured material, however, and have not been recorded by other authors. Also, they placed some of Riva's (1988) specimens identified as *Scalarigraptus* [= *Normalograptus*] *angustus* in their synonymy with *N. parvulus*. Riva (1988, p. 233) stated that most of this material was poorly preserved and distorted and for this reason alone I therefore follow Chen *et al.* (2005*a*, p. 266) in placing this material only questionably in *N. parvulus*.

Stratigraphical range. Zalasiewicz and Tunnicliff (1994, text-fig. 2) indicated that in central Wales *Normalograptus parvulus* is restricted to the *persculptus* Biozone and lower *acuminatus* Biozone (= lower *ascensus-acuminatus* Biozone). However, in the Wye Valley section (Zalasiewicz and Tunnicliff 1994, text-fig. 3) *N. parvulus* is shown to occur also in the middle and upper *acuminatus* Biozone. In Arctic Canada *N. parvulus* is recorded (as *Normalograptus* cf. *N. persculptus*) only from the upper *persculptus* Biozone (Melchin *et al.* 1991, p. 1856). The species has been recorded also from Dob's Linn, Moffat, southern Scotland (as *G.* cf. *persculptus*) from the *persculptus* and lower *acuminatus* biozones (Williams 1983; but note that Williams' investigations did not extend into the upper *acuminatus* Biozone) and from the *persculptus* Biozone of the upper Yangtze region of China (Chen *et al.* 2005*a*).

In Jordan the species ranges through to the top of the *ascensus-acuminatus* Biozone in the BG-14 core; thus its overall range is extended and would appear to be *persculptus* to *ascensus-acuminatus* biozones. However, the presence of *N. parvulus* in the section at Locality 7 (Text-fig. 4) only 0.24 m below strata containing *Rhaphidograptus toernquisti* (and *Cystograptus vesiculosus*) suggests that the range of *N. parvulus* may extend into the lower *vesiculosus* Biozone. In peri-Gondwanan Europe *R. toernquisti* appears some distance above the base of the *vesiculosus* Biozone (e.g. Štorch 1994).

Normalograptus persculptus (Elles and Wood, 1907)
Text-figure 18N

*. 1907   *Diplograptus (Glyptograptus) persculptus*, Salter; Elles and Wood, p. 257, pl. 31, fig. 7*a–c*; text-fig. 176*a–b*.

?p 1974   *Glyptograptus (G.) persculptus* (Salter, 1865); Hutt, p. 28 (*non* pl. 6, figs 9–12; = *N. parvulus*).

. 1975   *Glyptograptus persculptus* (Salter, 1865); Bjerreskov, p. 30, fig. 11A–C.

? 1978   *Glyptograptus? persculptus* (Slater) (*sic*); Baillie *et al.*, p. 46, fig. 2.

. 1983   *Glyptograptus persculptus* (Salter, 1865); Williams, p. 622, pl. 66, figs 1–3.

. 1984   *Glyptograptus? persculptus* (Salter, 1865); Vandenberg *et al.*, p. 10, figs 8A–D, 9A–B.

non 1987   *Glyptograptus persculptus* Salter; Wang, pl. 1, fig. 1.

1988   *Glyptograptus persculptus* (Salter); Cuerda *et al.*, p. 753, fig. 6f–g.

non 1994   *Normalograptus? persculptus* (Elles and Wood, 1907); Zalasiewicz and Tunnicliff, p. 704, text-fig. 5ᴀ–ᴄ.

non 1996   *Glyptograptus persculptus* (Salter); Rickards *et al.*, p. 110, figs 5a–d, 9a–b.

vp 1996   *Normalograptus persculptus* (Elles and Wood); Štorch and Loydell, p. 872, text-figs 3ᴀ–ᴅ, ꜰ–ʜ (*non* ᴇ = *parvulus*), 4–5 [see for full synonymy list; see also Chen *et al.* 2005*a*, pp. 264–265].

non 1998   *Persculptograptus persculptus* s.l. (Elles and Wood); Underwood *et al.*, p. 103, fig. 5S–W.

2001   "*Glyptograptus*" ("*Glyptograptus*") e.g. *persculptus* (Salter, 1865); Legrand, p. 150, text-fig. 8f.

p 2003   *Normalograptus* cf. *persculptus* (Elles and Wood); Masiak *et al.*, figs. 4a, c, 7a, (*non* 4b).

2005a   *Normalograptus persculptus* (Elles and Wood, 1907); Chen *et al.*, p. 266, text-fig. 9F, L (?5B, 9A).

Lectotype. Designated by Williams (1983, p. 622, pl. 66, fig. 3); BGS GSM 11782, from Ogofau, Pumpsaint, Wales.

Material. Two specimens, one preserved slightly obliquely as a very low-relief mould, from Locality 5, the other preserved flattened, from a depth of 3497 ft (*c.* 1066 m), core JF-1.

Description. The low-relief specimen is 7.5 mm long and bears six undamaged thecal pairs. Both the distalmost thecal pair and the proximal end are damaged. The rhabdosome increases in DVW from 1.2 to 1.95 mm over five thecal pairs. 2TRD is 1.85 mm throughout the length of the rhabdosome. The thecae have moderately sharp genicula, inclined, sometimes slightly convex supragenicular walls and sigmoidally curved interthecal septa. The thecal axes are inclined at *c.* 20 degrees to the rhabdosome axis. A median septum extends for much of the length of the rhabdosome and one of its margins extends, parallel to the nema, from the distal end of the rhabdosome. The flattened specimen is similar in lacking a proximal end; it bears seven thecal pairs. It attains a DVW of 1.8 mm with a 2TRD of 1.9 mm. A median septum is present, but in the absence of proximal ends, its completeness cannot be determined.

Remarks. This species was discussed at length by Štorch and Loydell (1996). Discussion here is limited to new observations.

One of the specimens illustrated by Zalasiewicz and Tunnicliff (1994, fig. 5B) as *N. persculptus* is identified herein as *N. parvulus* (see above). One of their figured specimens (fig. 5C) is questionably attributed to this taxon as it is only an early (possibly 'aberrant'; Zalasiewicz and Tunnicliff 1994, p. 704) growth stage. The other specimen (fig. 5A; BGS JZ 417F), stated to be from the lower *acuminatus* Biozone, but actually from the upper part of the biozone (see Zalasiewicz and Tunnicliff 1994, fig. 3), has its proximal thecae inclined at a higher angle (40–50 degrees) to the rhabdosome axis than those of *N. persculptus* (20–30 degrees). Of previously published taxa, the specimen resembles most closely *Normalograptus extraordinarius* (Sobolevskaya, 1974), for example those illustrated by Koren' *et al.* (1980, fig. 44) as '*Glyptograptus? persculptus* forma A', but it is narrower distally. It is placed in questionable synonymy with *Neodiplograptus* sp. 1 herein (see above).

N. persculptus is similar to *N. ojsuensis* (Koren' and Mikhaylova, *in* Apollonov *et al.* 1980). The latter differs in having sharper genicula proximally. This is clearly seen in the original material (Koren' and Mikhaylova 1980, pl. 41, figs 1–8; pl. 42, figs 1–2; text-fig. 43*a–e*) and in Legrand's (1993, fig. 2a–g) material from Niger.

Stratigraphical range. There appear to be no confirmed records of *N. persculptus* from the Silurian. The species is restricted to the Hirnantian *persculptus* Biozone. Melchin *et al.* (2003, p. 101) recorded *N. persculptus* from the '*extraordinarius* Band' at Dob's Linn, Moffat, southern Scotland.

Normalograptus rectangularis (McCoy, 1850)
Plate 1, figure 8; Text-figure 18H, R, T

1850 *Diplograpsus rectangularis* (McCoy); McCoy, p. 271.
. 1906 *Climacograptus rectangularis* (M'Coy); Elles and Wood, p. 187, pl. 26, fig. 5*a–b*, *e* (?5*c–d*); text-fig. 121*a* (?121*b*) [with additional synonymy].
1920 *Climacograptus rectangularis* M'Coy sp.; Gortani, p. 13, pl. 1, figs 11–12.
1920 *Climacograptus rectangularis* var. *alpinus* n. f. Gortani, p. 14, pl. 1, figs 13–15.
? 1948 *Climacograptus rectangularis* M'Coy; Wærn, pl. 26, fig. 8, text-fig. 5.
? 1965 *Climacograptus rectangularis* (M'Coy, 1850); Stein, p. 160, text-figs 15a–c, 16h.
1965 *Hedrograptus rectangularis* (M'Coy) 1850; Obut *et al.*, p. 29, pl. 1, figs 7–9.
1966 *Hedrograptus rectangularis* (M'Coy) 1850; Obut and Sobolevskaya, p. 10, pl. 3, fig. 4; text-fig. 3.

1967 *Hedrograptus rectangularis* (M'Coy) 1850; Obut *et al.*, p. 49, pl. 1, figs 10–12.
. 1970 *Climacograptus rectangularis* (M'Coy, 1850); Rickards, p. 30, pl. 3, fig. 1; text-fig 13, fig. 5.
. 1974 *Climacograptus rectangularis* (McCoy, 1850); Hutt, p. 19, pl. 1, figs 4–5.
. 1975 *Climacograptus rectangularis* (M'Coy, 1850); Bjerreskov, p. 24, pl. 4, fig. A.
1976 *Hedrograptus rectangularis* (McCoy, 1850); Sennikov, p. 135, pl. 5, figs 1–3.
1977 *Climacograptus* (*Climacograptus*) *rectangularis rectangularis* (F. M'Coy, 1850); Legrand, p. 149, text-fig. 1A–B.
1982 *Climacograptus rectangularis* (M'Coy); Lenz and McCracken, fig. 5d.
. 1988 *Climacograptus rectangularis* McCoy; Cuerda *et al.*, fig. 6h.
. 1988 *Climacograptus rectangularis* M'Coy; Rickards, fig. 1e.
1989 *Climacograptus rectangularis* (M'Coy); Melchin, fig. 7G.
1990 *Climacograptus medius* Tornquist; Fang *et al.*, p. 67, pl. 11, fig. 12 (*non* 5, 13–14).
1996 *Normalograptus rectangularis* (McCoy); Koren' and Rickards, pl. 5, figs 4–5, 7–8.
non 1996 *Climacograptus rectangularis* McCoy; Rickards *et al.*, p. 113, figs 5i–l, 9E, 10A–B.
non 1998 *Normalograptus rectangularis* (M'Coy, 1850); Melchin, text-fig. 3H–I.
1998 *Normalograptus rectangularis*; Underwood *et al.*, fig. 5K.
1998 *Normalograptus* aff. *rectangularis* (M'Coy); Underwood *et al.*, fig. 5N–O.
1999 *Normalograptus normalis* (Lapworth, 1877); Maletz, p. 344, fig. 2/15.
non 2003 *Normalograptus rectangularis* (M'Coy); Chen *et al.*, fig. 1K.
2003 *Normalograptus rectangularis* (M'Coy); Masiak *et al.*, fig. 8d (?5f).

Holotype. SM A20098a, figured by Elles and Wood (1907, pl. 26, fig. 5*a*); from the Birkhill Shales of Moffat, southern Scotland.

Material. 13 specimens: seven from core BG-14: depths 28.5 m (preserved in low relief), 32.5 m (one specimen preserved in low relief) and 33.5 m; one from Locality 7; three from Locality 10; two specimens, one on a bedding surface with *Parakidograptus acuminatus* (Locality 13).

Description. The sicula is 1.75–2.0 mm long ($n = 2$). Apertural width is 0.3 mm. The length of the exposed dorsal sicular margin is 0.25–0.4 mm ($n = 5$). The sicular apex reaches from the aperture of th2^1 to just above that of th2^2. The virgella is up to 5.5 mm long and variably robust. Th1^1 grows downwards for 0.1–0.4 mm below the sicular aperture. The thecae are of typical climacograptid form, with sharp genicula and supra-genicular walls very gently inclined to the rhabdosome axis proximally and parallel to it distally. The thecal apertures are horizontal. Measurements of DVW and 2TRD are given below. DVW is dependent upon preservation style, with relief material much narrower than flattened. There is a complete median septum.

DVW and 2TRD measurements (in mm)

th	1^1	2^1	3^1	5^1	10^1	15^1	20^1
DVW	0.6–1.05	0.75–1.1	1.0–1.3	1.3–1.55	1.5–1.8	1.75–1.9	2.0
2TRD		1.3–1.85	1.6–1.85	1.55–1.75	1.35–1.95	1.6–1.85	1.75–1.8

($n = 7–9$ for th1^1–5^1; $n = 5$ for th10^1; $n = 2$ or 3 for th15^1; and $n = 1$ or 2 for th20^1).

Remarks. N. rectangularis is a distinctive species, characterized by a narrow proximal end and then rapid widening. Distal width is quite variable, often exceeding 2 mm, but not in any of the Jordanian specimens described herein, which resemble most closely the specimen illustrated by Rickards (1988, fig. 1e). The virgella is often referred to as 'conspicuous' (e.g. Rickards 1970, p. 30) or 'robust' (e.g. Elles and Wood 1907, p. 188). Rickards *et al.* (1996) illustrated 'genicular flanges' on their material; these have not been recorded in N. rectangularis; hence, the rejection of this material in the synonymy list; it may belong in *Paraclimacograptus*. Melchin (1998) illustrated specimens described as 'spinose variants'; this spinosity (sicular and thecal) suggests that these specimens do not belong in N. rectangularis. Chen *et al.*'s (2003) specimen has inclined supragenicular walls and is rejected for this reason.

Stratigraphical range. N. rectangularis is a useful species as it does not occur in the lowermost Silurian. Toghill (1968), Rickards (1988, p. 346) and Masiak *et al.* (2003) recorded it from the upper *acuminatus* Biozone, from which it occurs also in Jordan. Examination of Toghill's collection in the Natural History Museum reveals the presence of N. rectangularis in the higher part of the interval yielding N. trifilis (Manck, 1923). All other records are from the *vesiculosus* Biozone through to lower Aeronian.

Normalograptus targuii Legrand, 2001
Plate 2, figure 4; Text-figure 18F–G, K, P–Q, S

*. 2001 *Normalograptus (Normalograptus) targuii* nov. sp. Legrand, p. 148, pl. 12, figs 11–12; text-fig. 7a–g.

Holotype. By original designation, the specimen figured by Legrand (2001, pl. 12, fig. 11; text-fig. 7a), from the In Azaoua region, Tassili-Oua-n-Ahaggar, Algeria.

Material. 22 specimens: 21 from core BG-14; one from Locality 7.

Description. The sicula is 1.5–2.0 mm long ($n = 10$). Apertural width is 0.25–0.35 mm. The length of the exposed dorsal sicular margin is 0.15–0.4 mm ($n = 13$; in eight specimens it is 0.25 mm). The sicular apex reaches the aperture of th2^2. The virgella is up to 9 mm long. Th1^1 grows downwards for 0.1–0.4 mm below the sicular aperture ($n = 14$; in 11 the distance is 0.15–0.25 mm). The thecae have sharp genicula and supragenicular walls consistently inclined at a low angle to the rhabdosome axis. The thecal apertures are horizontal. Measurements of DVW and 2TRD are given below. There is a complete median septum.

DVW and 2TRD measurements (in mm)

th	1^1	2^1	3^1	5^1	10^1
DVW	0.7–1.0	0.8–1.15	0.9–1.25	1.0–1.45	1.05–1.6
2TRD		1.3–1.8	1.4–1.75	1.4–1.85	1.45–2.1

($n = 19–21$ for th1^1–5^1; $n = 7$ or 8 for th10^1).

Remarks. N. targuii differs from the otherwise very similar N. ajjeri in the inclination of its supragenicular walls.

Stratigraphical range. The Jordanian specimens from core BG-14 are from the upper *ascensus-acuminatus* Biozone and from 1 m above the occurrence of *Parakidograptus acuminatus*. Legrand's (2001) material was from a geographically isolated exposure lacking stratigraphically diagnostic associated graptolites. The age was considered to be probably late Ordovician. In the BG-14 core N. targuii is common within its restricted stratigraphical range (Text-fig. 3). If this represents its total stratigraphical range then the age of Legrand's (2001) material is younger than he tentatively suggested.

Normalograptus transgrediens (Wærn, 1948)
Text-figure 16M

1929 *Climacograptus scalaris–C. medius* transient; Davies, pp. 8, 22, figs 28, 31.
*. 1948 *Climacograptus scalaris* His. v. *transgrediens* n. var. Wærn, p. 452, pl. 26, figs 2–3
? 1967 *Hedrograptus scalaris transgrediens* (Waern), 1948; Obut *et al.* p. 50, pl. 2, figs 1–2.
? 1978 *Climacograptus transgrediens* Waern; Chen and Lin, p. 30, pl. 4, fig. 22.
1982 *Climacograptus transgrediens* Wærn; Howe, pl. 1, figs a–b.
? 1990 *Climacograptus transgrediens* Waern; Fang *et al.*, p. 73, pl. 13, fig. 14.
2003 *N. transgrediens*; Koren' *et al.*, fig. 15.

Holotype. By original designation: gr. 1223b; figured Wærn (1948, pl. 26, fig. 2), from the *acuminatus* Biozone, Kullatorp core, Kinnekulle, Vestergötland, Sweden.

Material. One specimen, preserved in low relief, from a depth of 28.5 m, core BG-14.

Description. The specimen is a reverse view, damaged longitudinally distal to th5[1]. The rhabdosome tapers. Thecae are of typical climacograptid form, with sharp genicula and straight supragenicular walls parallel to the rhabdosome axis. Measurements of DVW and 2TRD are given below. The base of the median septum is just above the top of th2[2].

DVW and 2TRD measurements (in mm)

th	1[1]	2[1]	3[1]	5[1]
DVW	0.65	0.85	0.95	1.15
2TRD		1.45	1.55	1.5

($n = 1$).

Remarks. Wærn (1948) recognized four 'forma' of *N. transgrediens*, based upon the distance from the top of th1[1] to that of th5[1] and the level of the base of the median septum on the reverse side of the rhabdosome. The Jordanian specimen is a typical 'forma α'.

Stratigraphical range. Previous records of this species are from the upper *persculptus* Biozone (Davies 1929), post-*persculptus*-pre-*ascensus* interval (Koren' *et al.* 2003) through to the *acuminatus* Biozone (Davies 1929; Wærn 1948). Note that Wærn's (1948) specimens were, with one exception only, from below the single level in the core that yielded *Parakidograptus acuminatus*. Underwood *et al.* (1998, p. 97, but not illustrated) used *N. transgrediens* as an indicator of the *acuminatus* Biozone in the absence of the index species. The Jordanian specimen is from higher in the Rhuddanian, from the *vesiculosus* Biozone. Howe (1982) used the presence of *N. transgrediens* in the lower Solvik Formation of the Oslo region, Norway, to indicate the upper *persculptus* or *acuminatus* Biozone. The presence in Jordan of *N. transgrediens* in the *vesiculosus* Biozone indicates that the lower Solvik Formation may be younger than previously thought.

Normalograptus trifilis (Manck, 1923)
Text-figure 15K

p 1906 *Climacograptus medius*, Törnquist; Elles and Wood, p. 189, pl. 26, fig. 4*f* (*non* 4*a–e*); text-fig. 122*b* (*non* 122*a, c*).

*. 1923 *Climacograptus trifilis* spec. nov. Manck, p. 288, fig. 32.

. 1965 *Climacograptus trifilis* Manck, 1923; Stein, p. 165, pl. 14, fig. D; text-fig. 17a–d.

. 1975 *Climacograptus trifilis trifilis* Manck, 1923; Bjerreskov, p. 23, text-fig. 9B.

1982 *Climacograptus trifilis* Manck; Lenz and McCracken, fig. 4b.

. 1983 *Climacograptus trifilis* Manck, 1923; Williams, p. 618, text-fig. 5c.

. 1986 *Climacograptus trifilis* Manck; Štorch, pl. 5, fig. 5.

. 1988 *Climacograptus trifilis* Manck; Rickards, fig. 1r–s.

. 1990 *Climacograptus trifilis* Manck; Gnoli *et al.*, text-fig. 4a–b.

. 1991 *Normalograptus? trifilis*; Gutiérrez-Marco and Robardet, fig. 2d.

. 1993 *Normalograptus trifilis* (Manck, 1923); Štorch and Serpagli, p. 25, pl. 1, figs 2, 5; pl. 5, fig. 5; text-fig, 7G–J.

. 1995 *Normalograptus trifilis* (Manck); Piçarra *et al.*, fig. 3-1–2.

p 1996 *Normalograptus trifilis* (Manck); Koren' and Rickards, p. 38, pl. 5, figs 9, 11 (?10, *non* 12 = *N. lubricus* Chen and Lin 1978); text-fig. 23G.

. 1996 *Normalograptus trifilis trifilis* (Manck); Štorch, fig. 4.3a.

1999 *Normalograptus* (?) *trifilis* (Manck, 1923); Maletz, p. 346, figs 2/14, 4/8 (?figs 2/11, 3/4, 4/9).

Holotype. By monotypy; the specimen figured by Manck (1923, fig. 32), from the Llandovery of Thuringia.

Material. One proximal end, from a depth of 36.8 m, core BG-14.

Description. The single proximal end is broken obliquely across th3[2]–th4[1]. The proximal end is rounded. Sicular details are obscure, although three straight spines can be seen extending from the sicula, the longest being 0.7 mm long. DVW is 1.1 mm at th1[1] and 1.3 mm at th2[1]. 2TRD at th2[1] is 1.4 mm. Thecae are of typical climacograptid form.

Remarks. *N. trifilis* differs from *N. lubricus* (Chen and Lin, 1978) in being broader. Both taxa possess three straight spines, directed away from the sicular aperture and apparently arising at the base of the virgella. One of Maletz's (1999, fig. 3/4, 4/9) specimens may be *N. lubricus*. It is notably more slender than other figured material of *N. trifilis*. If it is *N. lubricus*, this would be the first record of this species from Avalonia. Štorch and Serpagli (1993, p. 26) discussed the two forms of sicular spinosity recorded in *N. trifilis* by Stein (1965).

Stratigraphical range. *Normalograptus trifilis* is very useful stratigraphically, being characteristic of the middle of the *ascensus-acuminatus* Biozone (Štorch 1996). It occurs rarely also in the upper part of the biozone, the level from which the single Jordanian specimen derives.

Normalograptus sp.
Text-figure 16I

Material. One proximal end, from a depth of 30.0 m, core BG-14.

Description. The proximal end is a reverse view and bears four thecal pairs. The rhabdosome is truncated by the edge of the core. The sicula appears to be narrow, 0.2 mm wide at its aperture, and 1.6 mm long, its apex reaching the apertures of the second thecal pair. The dorsal margin of the sicula is exposed for 0.2 mm. Th1[1] grows downwards for 0.2 mm below the sicular aperture. A narrow virgella extends 0.4 mm from the sicular aperture. The rhabdosome widens quite rapidly. DVW and 2TRD measurements are given below. The thecae are of glyptograptid form, with sigmoidally curved ventral walls and everted thecal apertures. The median septum appears to be complete.

DVW and 2TRD measurements (in mm)

th	1[1]	2[1]	3[1]	4[1]
DVW	0.85	1.15	1.4	1.5
2TRD		1.45	1.4	

($n = 1$).

Remarks. This specimen is not sufficiently complete to be assigned confidently to a species. It is clearly different from the other taxa described herein and is illustrated here for the sake of completeness.

Genus PARACLIMACOGRAPTUS Přibyl, 1947, emend. Russel *et al.*, 2000

Paraclimacograptus libycus (Desio, 1940)
Plate 2, figure 7; Text-figure 20C–H

*. 1940 *Climacograptus libycus* n. sp. Desio, p. 28, pl. 1, figs 1–8, 10–14.

. 1968 *Climacograptus innotatus jordaniensis* n. ssp. Wolfart *et al.*, p. 546, pl. 50, figs 1–4.

p 1976 *Climacograptus innotatus brasiliensis* Ruedemann; Jaeger, pl. 2, fig. 6; pl. 3, fig. 6 (*non* pl. 3, figs 1–3, 9–10).

1988 *Paraclimacograptus?* sp. nov.; Cuerda *et al.*, p. 753, fig. 5a–c.

? 1996 *Diplograptus* sp.; Rickards *et al.*, p. 108, figs 4e–f, 9D.

. 2002 *Normalograptus* (*Paraclimacograptus?*) *libycus* (Desio, 1940); Legrand, p. 219, pl. 12, figs 7–9; text-fig. 5l–n.

. 2004 *Paraclimacograptus? libycus* (Desio); Štorch and Massa, fig. 4.4–4.6.

. 2006 '*Paraclimacograptus*' *libycus* (Desio, 1940); Štorch and Massa, p. 96, pl. 1, figs 1–2, 5–6; text-fig. 7T–CC2 [with additional synonymy].

Lectotype. Designated by Štorch and Massa 2006, p. 96; housed at the Department of Earth Science, University of Milan; figured by Desio (1940, pl. 1, fig. 1), from the Tanezzuft Formation of the western Murzuq Basin, km 45 along the Serdeles–Ghat road, south-west Libya.

Material. 19 specimens, all from locality 6 (higher horizon).

Description. The apex of the sicula was visible in one specimen only. In this specimen, the sicular length is 1.4 mm, the apex reaching the level of the aperture of th2[1]. Its apertural width is 0.3 mm ($n = 3$). The length of the dorsal sicular margin exposed is 0.2–0.25 mm ($n = 4$). The virgella is short, less than 1 mm long. Th1[1] grows downwards for 0.3–0.35 mm below the sicular aperture ($n = 4$). The proximal end is conspicuously asymmetrical. The thecae have characteristically short supragenicular walls and correspondingly long apertures. Genicula are sharp; supragenicular walls are parallel to the rhabdosome axis or very gently inclined. Prominent, proximoventrally directed genicular flanges are visible on some thecae. Measurements of DVW and 2TRD are given below. A median septum is visible in some specimens; in one (Text-fig. 20G) it commences at the level of the aperture of th4[1], with pronounced undulation initially which diminishes distally.

DVW and 2TRD measurements (in mm)

th	1[1]	2[1]	3[1]	5[1]	10[1]	15[1]	20[1]
DVW	0.65–0.8	0.85–1.2	1.0–1.45	1.25–1.7	1.7–2.55	2.4–2.5	2.75
2TRD		1.45–1.9	1.45–1.85	1.45–1.85	1.6–1.95	1.55–1.85	1.5–1.6

($n = 6$ for th1[1]–10[1]; $n = 2$ for th15[1] and th20[1]).

Remarks. *Pa. libycus* is a distinctive species, characterized by an asymmetrical proximal end and very short supragenicular walls. Legrand (2002) and Štorch and Massa (2006) suggested that *Climacograptus innotatus jordaniensis* Wolfart *et al.*, 1968 is a junior synonym of *Pa. libycus*. This is confirmed herein: the Jordanian material is identical to the Libyan specimens described by Štorch and Massa (2006), who discussed *Pa. libycus* at length, noting that it differs from *Pa. brasiliensis* (Ruedemann, 1929) in having a tapering (rather than largely parallel-sided) and broader rhabdosome. The maximum DVW of *Pa. brasiliensis* is 2 mm.

Stratigraphical range. *Pa. libycus* appears to be confined to the Aeronian. In Jordan, it has been found with a triangulate monograptid (Wolfart *et al.* 1968), in Algeria in the lower Aeronian (Legrand 2002) and in Libya up to the *convolutus* Biozone (Štorch and Massa 2006).

Paraclimacograptus obesus (Churkin and Carter, 1970)
Text-figure 16K

p 1906 *Climacograptus innotatus*, Nicholson; Elles and Wood, p. 212, text-fig. 143*a* (*non* 143*b*, pl. 27, figs 10*a–e*).

* 1970 *Climacograptus innotatus obesus* n. subsp. Churkin and Carter, p. 16, pl. 1, fig. 3; text-fig. 6*D–E*.

1976 *Paraclimacograptus innotatus innotatus* (Nicholson, 1869); Sennikov, p. 122, pl. 3, figs 10–11.

1989 *Paraclimacograptus innotatus obesus* (Churkin and Carter); Melchin, fig. 5D.

1989 *Paraclimacograptus innotatus innotatus* (Nicholson); Melchin, fig. 8M.

p 1996 *Paraclimacograptus innotatus* (Nicholson, 1869); Koren' and Rickards, p. 49 (*non* pl. 8, figs 2–3) [see Russel *et al.* 2000, p. 90].

? 1998 *Paraclimacograptus innotatus innotatus* (Nicholson); Underwood *et al.*, fig. 5II.

. 2000 *Paraclimacograptus obesus*; Russel *et al.*, fig. 2.1.

p 2000 *Paraclimacograptus innotatus* (Nicholson, 1869); Koren' and Melchin, p. 1104, fig. 10.3–10.4, 10.8 (*non* fig. 10.5–10.7).

2003 *Paraclimacograptus innotatus* (Nicholson); Chen *et al.*, fig. 1M.

Holotype. By original designation, Churkin and Carter, 1970, p. 16; USNM 161611; from the upper Rhuddanian of Esquibel Island, Alaska.

Material. Three proximal ends, from a depth of 30.0 m, core BG-14.

Description. The three specimens are proximal ends (one damaged), bearing up to five thecal pairs. Details of the sicula are largely unclear; the length is *c.* 1.9 mm and the apex attains the level of the top of th2^2 in one specimen (Text-fig. 16K). The sicula has a virgella up to 0.75 mm long. The rhabdosome widens rapidly and thecae are closely spaced; details of DVW and 2TRD are given below. Thecae are of typical paraclimacograptid form, with sharp genicula, inclined supragenicular walls and conspicuous genicular flanges. The median septum appears to commence immediately distal of the sicular apex.

DVW and 2TRD measurements (in mm)

th	1^1	2^1	3^1	5^1
DVW	0.85	0.95–1.05	1.1–1.2	1.15–1.3
2TRD		1.2–1.3	1.15–1.35	

(*n* = 2).

Remarks. Pa. obesus differs from *Pa. innotatus* (Nicholson, 1869) and *Pa. exquisitus* (Rickards, 1970) in being more robust (see Russel *et al.* 2000, fig. 5, tables 1–2) and in having a longer sicula.

Stratigraphical range. Churkin and Carter (1970) recorded *Pa. obesus* only from the *cyphus* Biozone. Melchin (1989) recorded the species from the upper *cyphus* and the *acuminatus* biozones. The Jordanian material thus seems to be from the middle of this species' range, which appears to encompass much of the Rhuddanian.

Paraclimacograptus sp.
Plate 2, figure 8; Text-figure 10H

? 1998 *Pseudorthograptus (Pseudorthograptus) ?obuti* (Rickards and Koren'); Underwood *et al.*, p. 104, fig. 5HH.

Material. One proximal end from a depth of 30.0 m, core BG-14.

Description. The specimen is preserved in obverse view and bears two complete thecal pairs. The sicula is 1.9 mm long; the apertural margin is damaged dorsally. The sicular apex reaches the level of the second thecal pair. The dorsal margin of the sicula is exposed for 0.35 mm. Th1^1 grows downwards for 0.05 mm below the sicular aperture. DVW is 1.0 mm at th1^1 and 1.5 mm at th2^1. The ventral walls of the first thecal pair are inclined at 25 degrees to the rhabdosome axis giving the proximal end a distinctly triangular appearance. The first thecal pair have their dorsal walls slightly extended to produce hoods.

Remarks. Although only an early growth stage, the specimen is sufficiently different to justify it not being assigned to a described *Paraclimacograptus* species. The triangular proximal end is very similar to that of a specimen from Mauritania illustrated by Underwood *et al.* (1998) and tentatively identified by them as *Pseudorthograptus obuti* (Rickards and Koren', 1974). The latter species is very different, however, from the specimens of Underwood *et al.* in being ancorate and in having orthograptid thecae that are aperturally denticulate rather than possessing genicular processes. Underwood *et al.* (1998, fig. 3) placed their specimens in the lower *atavus* (= lower *vesiculosus*) Biozone based on very limited biostratigraphical data, but accepted (p. 97) that the level could be within the upper part of the *acuminatus* Biozone. It is interesting, stratigraphically, that the Jordanian specimen is from a level close to the *acuminatus/vesiculosus* biozone boundary.

The form of the proximal end bears some resemblance also to the stratigraphically older '*Paraclimacograptus*' *kiliani* (Legrand, 1977) and '*P.*' *kuramaensis* Koren' and Melchin, 2000, both of which are broader proximally.

Genus METACLIMACOGRAPTUS Bulman and Rickards, 1968

Metaclimacograptus hughesi (Nicholson, 1869)
Plate 2, figure 10; Text-figure 20A

*. 1869 *Diplograpsus Hughesi*, Nicholson, p. 235, pl. 11, figs 9–10.

p 1906 *Climacograptus Hughesi* (Nicholson); Elles and Wood, p. 208, pl. 27, fig. 11*a*, *d–e* (*non* 11*b–c*); text-fig. 140*a*, *c–d* (*non* 140*b*).

? 1965 *Climacograptus hughesi* (Nicholson, 1869); Stein, p. 167, text-fig. 14i.

1966 *Pseudoclimacograptus hughesi* (Nicholson), 1869; Obut and Sobolevskaya, p. 12, pl. 3, figs 5–6; text-fig. 4.

? 1967 *Pseudoclimacograptus hughesi* (Nicholson), 1869; Obut *et al.*, p. 52, pl. 2, fig. 4.

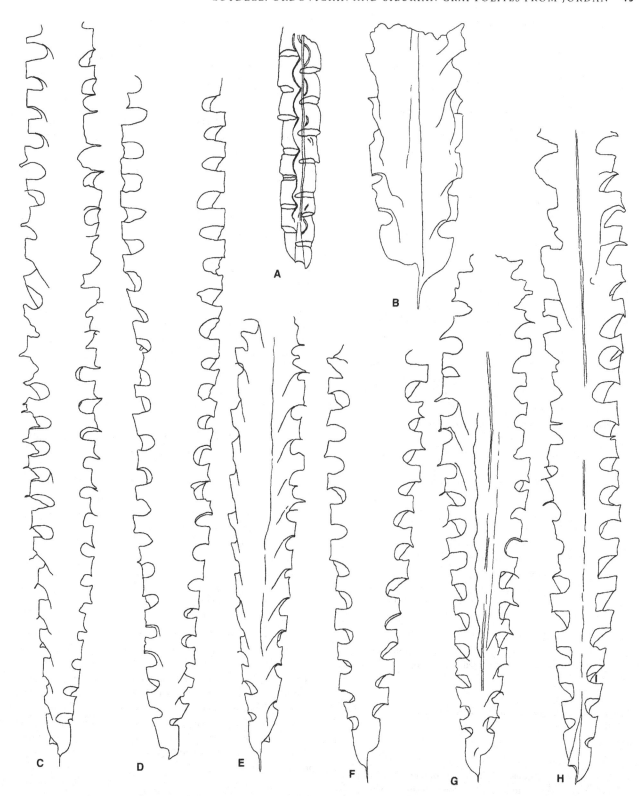

TEXT-FIG. 20. A, *Metaclimacograptus hughesi* (Nicholson, 1869), BGS FOR 5363b, core BG-14, 27.5 m, *vesiculosus* Biozone. B, *Cystograptus vesiculosus* (Nicholson, 1868b), GD-NRA 18401-54c, Locality 7, *vesiculosus* Biozone. C–H, *Paraclimacograptus libycus* (Desio, 1940), Locality 6, higher horizon (Aeronian). C, NHM QQ230. D, NHM QQ233. E, NHM QQ231(3). F, NHM QQ231(4). G, NHM QQ231(1). H, NHM QQ231(2). All figures × 10.

non 1968 *P. (Metaclimacograptus) hughesi* (Nicholson);
Bulman and Rickards, p. 3, text-fig. 1*a–c* [=
Metaclimacograptus slalom Zalasiewicz, 1996].

1970 *P. (Metaclimacograptus) hughesi* (Nicholson);
Churkin and Carter, p. 20, pl. 1, fig. 16 (?17);
text-fig. 8*E*.

?p 1970 *Pseudoclimacograptus (Metaclimacograptus)
hughesi* (Nicholson, 1869); Rickards, p. 33 (*non*
text-fig. 14, fig. 6).

non 1970 *P. (Metaclimacograptus) hughesi* (Nicholson);
Hutt *et al.*, p. 4, pl. 1, figs 1–4.

non 1974 *Pseudoclimacograptus (Metaclimacograptus)
hughesi* (Nicholson, 1869); Hutt, p. 22, pl. 2,
figs 6–7, 13–14.

p 1975 *Pseudoclimacograptus undulatus* (Kurck, 1882);
Bjerreskov, p. 26, pl. 4, fig. E.

non 1976 *Pseudoclimacograptus (Metaclimacograptus)
hughesi* (Nicholson, 1869); Sennikov, p. 124,
pl. 4, figs 1–3.

? 1982 *Pseudoclimacograptus (Metaclimacograptus) hughesi*
(Nicholson); Lenz and McCracken, fig. 5c.

. 1989 *Metaclimacograptus hughesi* (Nicholson); Melchin,
fig. 7I.

non 1990 *Pseudoclimacograptus (Undulograptus) hughesi*
(Nicholson); Chen and Qian, p. 2, pl. 1, figs 1–6,
10–14; pl. 2, figs 4–5, 8–11; pl. 3, figs 2–4, 7, 9;
text-fig. 1.

non 1991 *Metaclimacograptus hughesi* (Nicholson, 1869);
Loydell, p. 675, pl. 1, figs 3–4, 6, 9, 12.

. 1996 *Metaclimacograptus hughesi* (Nicholson, 1869);
Zalasiewicz, p. 2, text-fig. 2A–C.

. 1996 *Metaclimacograptus hughesi* (Nicholson, 1869);
Koren' and Rickards, p. 95, pl. 13, figs 9–15;
text-fig. 23A.

1998 *Metaclimacograptus hughesi* (Nicholson);
Underwood *et al.*, fig. 5X.

v. 2003 *Metaclimacograptus hughesi* (Nicholson); Loydell
et al., fig. 4d.

. 2006 *Metaclimacograptus hughesi* (Nicholson, 1869);
Štorch and Massa, p. 93, text-fig. 7N1–O2.

Neotype. Selected by Přibyl 1948, p. 18: NHM Q1365, figured by Elles and Wood (1906, pl. 27, fig. 11*a*) and Zalasiewicz (1996, text-fig. 2C), from the 'Skelgill Beds' of Ambleside, Cumbria, England.

Material. Six specimens, four from core BG-14, depth 27.5 m; two from Locality 7.

Description. The rhabdosome is up to 11 mm long with a rounded proximal end. Sicular details are obscure. In one specimen, the sicula appears to be 1.5 mm long, its apex reaching the aperture of th2^1. The thecae are of typical metaclimacograptid morphology with slit-like horizontal to introverted apertures and supragenicular walls inclined slightly inwards towards the succeeding thecal aperture. Measurements of DVW and 2TRD are given below. The median septum exhibits angular undulation. The nema is broad, up to 0.15 mm wide.

DVW and 2TRD measurements (in mm)

th	1^1	2^1	3^1	5^1	10^1	15^1
DVW	0.75–0.8	0.8–0.95	0.9–0.95	1.05	1.0–1.2	1.1
2TRD		1.25–1.4	1.25–1.5	1.4–1.7	1.5–1.65	1.45

($n = 4$ for th1^1–5^1; $n = 3$ for th10^1; $n = 1$ for th15^1).

Remarks. The material agrees well with Zalasiewicz's (1996) description of the neotype. As he noted, Bulman and Rickards (1968) and therefore subsequent authors who used this paper for identification purposes, misidentified *Me. hughesi*, assigning narrower material with a wavy rather than angular median septum to this taxon. Zalasiewicz (1996, p. 11) erected *Me. slalom* for Bulman and Rickards' (1968) material.

Stratigraphical range. Zalasiewicz (1996) recorded a range of *leptotheca* to lower *convolutus* biozones (middle Aeronian) for *Me. hughesi*. This range was extended upwards, into the *sedgwickii* Biozone, by Štorch and Massa (2006), and considerably downwards by Koren' and Rickards (1996), whose oldest material was from the *vesiculosus* Biozone, which is the age of the Jordanian material described here.

Genus SUDBURIGRAPTUS Koren' and Rickards, 1996

Sudburigraptus illustris (Koren' and Mikhaylova, 1980)
Text-figure 11D

*. 1980 *Orthograptus illustris* Koren' et Mikhaylova, sp. nov.,
p. 158, pl. 51, figs 1–6; text-fig. 50a–d.

2000 *Sudburigraptus illustris* (Koren' and Mikhaylova
1980) new combination; Koren' and Melchin,
p. 1109, figs 9.18, 10.1–10.2.

Holotype. By original designation, the specimen illustrated by Koren' *et al.* (1980, pl. 51, fig. 1; text-fig. 50a–b), from the *acuminatus* Biozone of Kazakhstan.

Material. Three specimens, two from a depth of 30.0 m, core BG-14; one from a depth of 3490′ (*c.* 1064 m), core JF-1.

Description. The sicula is 1.4 mm long in one specimen and is at least 1.6 mm long in another; its apex reaches the top of th2^1. The sicular apertural width is 0.3 mm. Measurements of DVW and 2TRD are given below. The first three thecal pairs exhibit sigmoidal curvature of their ventral walls; more distal thecae are orthograptid. A complete median septum is present.

DVW and 2TRD measurements (in mm)

th	1^1	2^1	3^1	5^1
DVW	0.95–1.15	1.3	1.35–1.4	1.4–1.8
2TRD		1.25–1.55	1.3–1.7	

($n = 2$).

Remarks. The Jordanian specimens match previous descriptions of this species: particularly characteristic is the transition proximally from glyptograptid to orthograptid thecae.

Stratigraphical range. This rare species has been recorded previously from the *acuminatus* Biozone of Kazakhstan (Koren' *et al.* 1980) and the *ascensus* Biozone of Uzbekistan (Koren' and Melchin 2000). Its range is thus extended herein to the *vesiculosus* Biozone.

Sudburigraptus sp.
Text-figures 12H, 14F

? 1989 *Orthograptus* n. sp.; Melchin, fig. 6E–F.

Material. Three specimens, two proximal ends and one distal fragment, from a depth of 30.0 m, core BG-14.

Description. The proximal ends are indifferently preserved, but the sicula can be seen in one specimen to be 2.1 mm long, reaching the level of the aperture of th2^2. There is a robust virgella, at least 2 mm long. The proximal end of the rhabdosome tapers conspicuously. Proximal thecae have a slight sigmoidal curvature to the ventral wall; distally they are more orthograptid. Thecal apertures are concave and everted. Measurements of DVW and 2TRD are given below. There is a complete median septum.

DVW and 2TRD measurements (in mm)

th	1^1	2^1	3^1	5^1	10^1	20^1
DVW	0.75–1.0	1.15–1.25	1.2–1.35	1.55–1.6	1.6	2.15
2TRD		1.45–1.7	1.55–1.8	1.75–2.0	1.95	2.1

($n = 2$ up to th5^1; $n = 1$ distally).

Remarks. This material is similar to '*Orthograptus*' *cabanensis* Zalasiewicz and Tunnicliff, 1994 (assigned tentatively to *Sudburigraptus* by Koren' and Rickards 1996, p. 49), but the rhabdosome of *S.* sp. is more strongly tapering proximally and th1^1–1^2 do not have concave ventral walls, as shown in Zalasiewicz and Tunnicliff's (1994) text-figure 9I–L.

Sudburigraptus sp. may be conspecific with Melchin's (1989) '*Orthograptus* n. sp.' from a similar stratigraphical level in arctic Canada: DVWs are similar and both possess tapering proximal ends. The 2TRD of the Canadian material is rather lower, however. '*Orthograptus truncatus*' *sensu* Hutt (1974) has more closely spaced thecae (Hutt 1974, p. 33 quoted 15 thecae in 10 mm).

Genus CYSTOGRAPTUS Hundt, 1942, emend.
Jones and Rickards, 1967

Cystograptus vesiculosus (Nicholson, 1868*b*)
Text-figure 20B

*. 1868*b* *Diplograpsus vesiculosus*, n. sp. Nicholson, p. 57, pl. 3, fig. 11.

. 1869 *Diplograpsus vesiculosus*, Nich.; Nicholson, p. 237, pl. 11, figs 14–15.

. 1907 *Diplograptus (Orthograptus) vesiculosus*, Nich.; Elles and Wood, p. 229, pl. 28, figs 8*a–d*; text-fig. 151*a–f*.

1965 *Cystograptus vesiculosus* (Nicholson), 1868; Obut *et al.*, p. 34, pl. 2, figs 1–4.

. 1967 *Cystograptus vesiculosus* (Nicholson); Jones and Rickards, p. 180, figs 3d, 6, 8.

1967 *Cystograptus vesiculosus* (Nicholson), 1868; Obut *et al.*, p. 63, pl. 3, figs 7–8; pl. 4, figs 1–11.

1968 *Cystograptus vesiculosus* (Nicholson), 1868; Obut *et al.*, p. 34, pl. 2, figs 1–4.

. 1970 *Orthograptus vesiculosus* (Nicholson); Churkin and Carter, p. 30, pl. 12, figs 11–13.

?p 1970 *Cystograptus vesiculosus* (Nicholson, 1868); Rickards, p. 44, ?pl. 1, fig. 11 (*non* pl. 2, fig. 14) [with Hundt synonymy entries].

p 1974 *Cystograptus vesiculosus* (Nicholson, 1868); Hutt, p. 45, pl. 5, figs 4–5 (*non* pl. 4, fig. 15; text-fig. 9, figs 4–5).

? 1975 *Cystograptus vesiculosus* (Nicholson, 1868); Bjerreskov, p. 29, fig. 10E.

1976 *Orthograptus (Cystograptus) vesiculosus* (Nicholson); Jaeger, pl. 3, fig. 7.

. 1978 *Orthograptus vesiculosus* (Nicholson); Chen and Lin, p. 41, pl. 7, figs 22–24.

1979 *Diplograptus (Cystograptus) vesiculosus* (Nicholson); Jaeger and Robardet, pl. 2, fig. 17.

? 1979 *Cystograptus vesiculosus* (Nicholson, 1868); Paškevičius, p. 128, pl. 4, figs 1–6; pl. 19, figs 18–20.

1982 *Cystograptus vesiculosus* Nicholson; Lenz and McCracken, fig. 5i.

1984 *Orthograptus vesiculosus* (Nicholson); Mu and Lin, p. 61, pl. 6, figs 1–2.

p 1985 *Cystograptus vesiculosus* (Nicholson, 1868); Štorch, p. 96, pl. 2, figs 5, 7 (*non* 1); text-fig. 3E–F (*non* G).

1986 *Cystograptus vesiculosus* (Nicholson); Štorch, pl. 6, fig. 5 (?2).

1987 *Orthograptus vesiculosus* (Nicholson); Wang, pl. 1, fig. 5.

. 1988 *Cystograptus vesiculosus* (Nicholson); Rickards, fig. 1q.

1990 *Orthograptus vesiculosus* (Nicholson); Fang *et al.*, p. 81, pl. 14, figs 3, 5; pl. 16, fig. 5.

. 1991 *Cystograptus vesiculosus*; Gutierrez-Marco and Robardet, fig. 2n–o.

. 1993 *Cystograptus vesiculosus* (Nicholson, 1868); Štorch and Serpagli, p. 14, pl. 5, fig. 3, text-fig. 5I.

. 1999 *Cystograptus vesiculosus* (Nicholson, 1868); Gutiérrez-Marco *et al.*, pl. 1, figs 1–3, 13z.

. 1999 *Cystograptus vesiculosus* (Nicholson, 1868); Maletz, p. 350, figs 2/13, 4/11.

?p 1996 *Cystograptus vesiculosus* (Nicholson, 1868); Koren' and Rickards, p. 32, pl. 3, figs 4–6, 8–10 (*non* fig. 7).

? 1998 *Cystograptus vesiculosus* (Nicholson, 1868); Melchin, text-fig. 6L–M, o.

p 2003 *Cystograptus vesiculosus* (Nicholson); Masiak *et al.*, fig. 5j, l (*non* fig. 5k).

2004 *Cystograptus vesiculosus* (Nicholson); Štorch and Massa, fig. 3.3.

non 2004 *Cystograptus vesiculosus* (Nicholson) *s.l.*; Koren' and Rickards, p. 885, text-figs 12–13.

Holotype. By monotypy, the specimen figured by Nicholson (1868b, pl. 3, fig. 11), from the Llandovery of Dob's Linn, Moffat, southern Scotland. Strachan (1997, p. 59) followed Benton (1979) in suggesting that the specimen may be NHM P1880.

Material. One proximal end and several other fragments, from Locality 7.

Description. Only the proximal end is well enough preserved to warrant description. The other material is preserved in what Elles and Wood (1907, p. 230) termed the characteristic 'pseudo-"bi-scalariform" view'. The proximal end is broad with a DVW of 2.0 mm at th1^1 and 2.3 mm at th2^1. 2TRD at th2^1 is 2.2 mm.

Remarks. Maletz (1999, p. 350) suggested that *Cystograptus vesiculosus* and *C. penna* (Hopkinson, 1869) might be synonyms, a view expressed also by Hutt (1974, p. 46: 'extreme variants of one species'). The two species are very different, however, in terms of their maximum DVW. *C. penna* attains a distal DVW of only 1.8 mm (in relief material; Jones and Rickards 1967, p. 174), whereas *C. vesiculosus* exceeds 3 mm in width distally. Also, it appears that the proximal thecae of *C. vesiculosus* are of significantly simpler morphology than those of *C. penna*. This was originally suggested by Jones and Rickards (1967, p. 180) and is supported by the Jordanian proximal end, by Rickards (1988, fig. 1q) and by Maletz's (1999) partial relief proximal end of *C. vesiculosus*. Elles and Wood (1907, p. 230) noted considerable variation in virgellar length, with a maximum recorded of 13 mm.

Stratigraphical distribution. The first appearance of *C. vesiculosus* marks the base of the *vesiculosus* Biozone. The highest recorded specimens are from the lower *cyphus* Biozone (Hutt 1974; Štorch 1985).

Genus AKIDOGRAPTUS Davies, 1929

Akidograptus ascensus Davies, 1929
Text-figure 21B–E

*. 1929 *Akidograptus ascensus* sp. nov. Davies, p. 9, figs 22–24.

. 1962 *Akidograptus ascensus* Davies; Tomczyk, pl. 4, figs 1–2; pl. 7, figs 1–2.

1964 *Akidograptus ascensus* Davies; Yang, p. 631, pl. 1, figs 6–11; text-fig. 2a–b.

1965 *Diplograptus* (*Akidograptus*) *ascensus* Davies, 1929; Stein, p. 176, pl. 14, fig. F; pl. 15, figs a–b; text-figs 22a, d, 23a–c.

? 1967 *Akidograptus ascensus* Davies, 1929; Obut *et al.*, p. 73, fig. 6, figs 8–9.

. 1974 *Akidograptus ascensus* Davies, 1929; Hutt, p. 55, text-fig. 9, figs 9–10.

p 1975 *Akidograptus ascensus* Davies, 1929; Bjerreskov, p. 42, fig. 13D (*non* 13E).

. 1976 *Akidograptus ascensus* Davies; Jaeger, pl. 2, fig. 4.

1978 *Akidograptus ascensus* Davies; Chen and Lin, p. 47, text-fig. 6d.

1979 *Akidograptus ascensus* Davies; Jaeger and Robardet, pl. 2, fig. 16; text-fig. 9a, d.

1981 *Akidograptus ascensus* Davies; Li and Ge, p. 227, pl. 1, figs 1–2.

. 1983b *Akidograptus ascensus* Davies, 1929; Štorch, p. 297, pl. 2, figs 5–9; text-fig. 1B–C, F, H.

. 1983 *Akidograptus ascensus* Davies, 1929; Williams, p. 629, text-figs 9f–g, i, 10j–m (?n).

1984 *Akidograptus ascensus* Davies; Lin and Chen, p. 217, pl. 6, fig. 6.

. 1986 *Akidograptus ascensus* Davies; Štorch, pl. 5, fig. 1.

1987 *Akidograptus ascensus* Davies; Wang, pl. 1, fig. 3.

1988 *Akidograptus ascensus* Davies; Rickards, fig. 1g–h.

1990 *Akidograptus ascensus* Davies; Fang *et al.*, p. 94, pl. 21, figs 6, 9; pl. 23, fig. 3; pl. 24, fig. 1.

. 1991 *Akidograptus ascensus*; Melchin and Mitchell, fig. 2.7.

. 1993 *Akidograptus ascensus* Davies, 1929; Štorch and Serpagli, p. 28, pl. 4, figs 2, 5; text-fig. 8C–D.

. 1994 *Akidograptus ascensus* Davies; Lenz and Vaughan, fig. 3J, P.

1995 *Akidograptus ascensus* Davies, 1929; Koren' and Rickards, p. 90, pl. 13, fig. 1; text-figs 21A–B, 22A–H.

. 1996 *Akidograptus ascensus* Davies; Štorch, fig. 4.5.

p 1998 *Akidograptus ascensus* Davies, 1929; Melchin, text-fig. 3s (*non* 3u).

. 1999 *Akidograptus ascensus* Davies, 1929; Maletz, p. 351, figs 2/3–2/4, 2/7, 3/2, 4/1–4/2, 4/6.

. 2000 *Akidograptus ascensus* Davies, 1929; Koren' and Melchin, p. 1109, figs 12.1–12.3, 13.1–13.2.

. 2000 *Akidograptus ascensus* Davis [*sic*], 1929; Rigby, fig. 1.2.

2003 *Akidograptus ascensus* (Davies); Chen *et al.*, fig. 1I.

. 2003   *A. ascensus*; Koren' et al., fig. 3.24.

2003   *Akidograptus ascensus* Davies; Masiak et al., figs 4i–j, p, 7i.

Holotype. By original designation; SM A10021, figured by Davies (1929, fig. 23) and Rigby (2000, fig. 1.2); from the Birkhill Shales of Dob's Linn, Moffat, southern Scotland.

Material. 17 specimens, all from core WS-6.

Description. The rhabdosome is thorn-shaped. The virgella branches 0.15–0.2 mm below the sicular aperture in thecate rhabdosomes, up to 0.5 mm from the sicular aperture in siculae assigned to the species. In some specimens (e.g. Text-fig. 21B–C) there appear to be primary and secondary bifurcations. The distance from the sicular aperture to the aperture of th1^1 is 2.65 and 2.7 mm in specimens from 1399.8 m, 2.9 mm in a specimen from 1399.45 m and 1.6 mm in one from 1398.3 m. The thecae have straight supragenicular walls, inclined to the rhabdosome axis at 5–10 degrees. Thecal apertures are of climacograptid form, horizontal and with sharp genicula. Details of proximal DVW and 2TRD are given below. There is a complete median septum.

DVW and 2TRD measurements (in mm)

th	1^1	2^1	3^1
DVW	0.5–0.6	0.7–0.8	0.8–1.0
2TRD		1.75–2.25	1.9–1.95

($n = 4$).

Remarks. Several new species of *Akidograptus* have been described from China (e.g. Yang 1964; Li and Ge 1981; Yu et al. 1988). These have not yet been recognized elsewhere. *A. ascensus* differs from *A. cuneatus* Chaletskaya, 1960 in being broader (maximum DVWs quoted by Koren' and Melchin 2000 are 1.0 mm and 0.68 mm, respectively).

One of Bjerreskov's (1975, fig. 13E) specimens appears to lack the apertural excavations characteristic of *A. ascensus* and is therefore excluded from this taxon (see also Vandenberg et al. 1984, p. 13). It resembles most *Parakidograptus angustitubus* Li, *in* Li and Ge 1981. If this identification is correct, this would be the first record of this species outside China.

Štorch (1983b) noted significant changes in *A. ascensus* rhabdosomes through the Běleč section, Bohemia: stratigraphically younger specimens possessed shorter siculae, lower 2TRDs and attained greater rhabdosome lengths. The type specimen has a 2TRD at th2^1 of 2.6 mm, significantly higher than the Jordanian specimens described herein. The distance from the sicular aperture to that of the aperture of th1^1 is clearly highly variable, both within the limited Jordanian material available and as measured on specimens described in the literature, ranging from 1.5 mm (Koren' *et al.* 2003, fig. 3.24) to

3.15 mm (Štorch 1983b, fig. 1B). Whether there is a stratigraphically useful trend in this character remains to be established.

It is clear that the ancora-like structure was developed very early in astogeny, prior to the origin of the thecae (see Text-fig. 21B; Stein 1965, fig. 23a–c, showed 'juvenile' specimens with th1^1 partially developed, also showing the presence of this ancora-like structure; see also Koren' and Rickards 1996, text-fig. 22A–F). This early development is the same as that demonstrated by Bates and Kirk (1984) in true ancorae from Aeronian diplograptoids.

Genus PARAKIDOGRAPTUS Li and Ge, 1981

Parakidograptus acuminatus (Nicholson, 1867)
Plate 1, figure 6; Text-figure 21F–G

*. 1867   *Diplograpsus accuminatus*, n. sp. Nicholson, p. 109, pl. 7, fig. 16 (?17).

. 1908   *Cephalograptus* (?) *acuminatus* (Nicholson); Elles and Wood, p. 289, pl. 32, fig. 11a–d; text-fig. 199.

1962   *Akidograptus acuminatus* (Nicholson); Tomczyk, pl. 4, figs 3–4; pl. 7, figs 3–5.

1965   *Diplograptus (Akidograptus) acuminatus acuminatus* (Nicholson, 1867); Stein, p. 174, pl. 15, fig. C; text-fig. 22e–f.

1967   *Akidograptus acuminatus* (Nicholson), 1867; Obut et al., p. 74, pl. 6, figs 10–13.

1970   *Akidograptus acuminatus* (Nicholson); Churkin and Carter, p. 34, pl. 3, figs 16–17; text-fig. 13B–C.

p 1974   *Orthograptus*? *acuminatus acuminatus* (Nicholson, 1867); Hutt, p. 37, pl. 7, fig. 9; text-fig. 10, fig. 4 (*non* text-fig. 9, fig. 11).

1976   *Akidograptus acuminatus* (Nicholson); Jaeger, pl. 3, fig. 4.

1979   *Akidograptus acuminatus* (Nicholson); Jaeger and Robardet, text-fig. 8c–d.

. 1981   *Parakidograptus acuminatus* (Nicholson); Li and Ge, p. 229, pl. 1, figs 8–10.

1982   *Orthograptus acuminatus* (Nicholson); Lenz and McCracken, fig. 4i (?l–m).

. 1983b   *Parakidograptus acuminatus* (Nicholson, 1867); Štorch, p. 298, pl. 1, figs 1–8; pl. 2, figs 1–4; text-fig. 1A, D–E, G, I–L.

. 1984   *Parakidograptus acuminatus* (Nicholson); Lin and Chen, p. 219, pl. 6, figs 7–8.

. 1984   *Parakidograptus acuminatus* (Nicholson); Mu and Lin, p. 63, pl. 6, fig. 10.

. 1986   *Parakidograptus acuminatus* (Nicholson); Štorch, pl. 5, fig. 3.

1987   *Parakidograptus acuminatus* (Nicholson); Wang, pl. 1, fig. 2.

1988   *Parakidograptus acuminatus* (Nicholson); Rickards, fig. 1j.

1989   *Parakidograptus acuminatus acuminatus*
       (Nicholson); Melchin, fig. 5F.

1990   *Parakidograptus acuminatus* (Nicholson);
       Fang *et al.*, p. 89, pl. 20, fig. 12; pl. 21, figs 3, 8;
       pl. 24, figs 6, 11; pl. 25, figs 1–3.

. 1991   *Parakidograptus acuminatus*; Gutiérrez-Marco
          and Robardet, fig. 2a–c.

. 1993   *Parakidograptus acuminatus* (Nicholson, 1867);
          Štorch and Serpagli, p. 28, pl. 4, figs 3, 6–7; text-
          fig. 8A–B.

. 1994   *Parakidograptus acuminatus* (Nicholson); Lenz
          and Vaughan, fig. 3K–M.

. 1995   *Parakidograptus acuminatus* (Nicholson); Piçarra
          et al., fig. 3-3–4.

1996   *Parakidograptus* cf. *acuminatus* (Nicholson,
       1867); Koren' and Rickards, p. 90, text-fig. 22I.

. 1996   *Parakidograptus acuminatus* (Nicholson); Štorch,
          fig. 4.6.

? 1998   *Parakidograptus acuminatus* (Nicholson, 1867);
          Melchin, text-fig. 3v.

. 1999   *Parakidograptus acuminatus* (Nicholson, 1867);
          Maletz, p. 352, figs 2/1, 2/5–6, 4/3.

non 2000   *Parakidograptus acuminatus acuminatus*
            (Nicholson, 1867); Koren' and Melchin, p. 1111,
            figs 12.7–12.14, 13.3, 13.7–13.8.

? 2003   *Parakidograptus acuminatus* (Nicholson); Chen
          et al., fig. 1L.

. 2003   *P. acuminatus*; Koren' *et al.* 2003, fig. 3.30–3.31.

. 2003   *Parakidograptus acuminatus* (Nicholson); Masiak
          et al., figs 4r, 5a, 7c, 8a, g.

2003   *Parakidograptus acuminatus* (Nicholson); Rong
       et al., fig. 4.16.

v. 2005   *Parakidograptus acuminatus* (Nicholson, 1867);
          Lüning *et al.*, fig. 15a.

Lectotype. The specimen illustrated by Nicholson (1867, pl. 7, fig. 16), from southern Scotland. Strachan (1997, p. 85) suggested that this may be NHM 1310.

Material. 14 specimens, six from core BG-14, depth 36.0 m; eight from Locality 13.

Description. The rhabdosome is up to 25 mm long and is thorn-shaped. It is either straight (Text-fig. 21F) or curved (Text-fig. 21G). The virgella branches up to 0.2 mm from the sicular aperture. The distance from the sicular aperture to the aperture of th1^1 is 2.4–2.5 mm in the core material and 2.2–2.55 mm in that from Locality 13. Th1^1 originates 0.45–0.55 mm above the sicular aperture in the core material and 0.7 mm above in the single specimen from Locality 13 in which this feature is visible. Proximal thecae resemble those of *Akidograptus ascensus* and have pronounced thecal excavations; from the second or third thecal pair the thecae are of orthograptid form. Apertures can be clearly seen to undulate in some thecae. The apertures are generally normal to the rhabdosome axis. A slight excavation is visible even in some distal thecae (Text-fig. 21G). Measurements of DVW and 2TRD are given below. There appears to be a complete median septum.

DVW and 2TRD measurements (in mm)

th	1^1	2^1	3^1	5^1	10^1	15^1
DVW	0.55–0.6	0.75–0.95	1.0–1.2	1.3–1.6	1.6–2.05	1.8
2TRD		2.1–2.4	2.0–2.05	1.9–2.15	2.2	1.9

($n = 5$ for th1^1–3^1; $n = 3$ for th10^1; $n = 1$ for th15^1).

Remarks. Several new species of *Parakidograptus* have been described from China (e.g. by Li and Ge 1981; Yu *et al.* 1988; Fang *et al.* 1990). These have generally not been recognized elsewhere (but see above, under 'Remarks' for *Akidograptus ascensus* and Masiak *et al.*'s 2003, figs 4k, 7d record of *P. primarius* Li, 1990).

Štorch (1983b) noted significant changes in *P. acuminatus* rhabdosomes through the Prague-Řepy section, with the stratigraphically older specimens being shorter with an extremely protracted proximal end, whilst the stratigraphically youngest specimens had lower 2TRDs, a more robust proximal end and much longer rhabdosomes (commonly 30–40 mm). The protraction of the proximal end can be quantified by measuring the distance from the sicular aperture to the aperture of th1^1. The following measurements were taken from specimens known to be from the lower part of the *acuminatus* Biozone: 4.1 mm (Hutt 1974, text-fig. 10, fig. 4), 3.2 and 3.4 mm (Štorch 1983b, fig. 1J, M, respectively), 2.85 mm (Maletz 1999, fig. 2/5; note that this is not the stratigraphically lowest occurrence), 3.5 and 3.7 mm (Koren' *et al.* 2003, figs 3.30–3.31; specimens from depth of 50.50–50.40 m; T. Koren', pers. comm. 2005). These figures are greater than for specimens from the upper part of the biozone: 1.8 mm (Štorch 1983b, fig. 1L), 2.2 mm (Maletz 1999, fig. 2/1). The Jordanian specimens described herein are also from the upper part of the biozone: the distance from the sicular aperture to the aperture of th1^1 in these is 2.4–2.5 mm.

Koren' and Melchin's (2000, p. 1111) specimens from the *Hirsutograptus sinitzini* Subzone of Uzbekistan differ from *P. acuminatus* in two respects: proximal 2TRD is low, with all but one figured specimen having a 2TRD at th2^1 of less than 2.0 mm (range is 1.4–2.2 mm); and, as Koren' and Melchin noted, the origin of th1^1 is lower on the sicula (they quoted 0.35 mm) than is typical of *P. acuminatus*.

Genus RHAPHIDOGRAPTUS Bulman, 1936

Rhaphidograptus toernquisti (Elles and Wood, 1906)
Text-figure 21J

*. 1906   *Climacograptus Törnquisti*, sp. nov. Elles and
          Wood, p. 190, pl. 26, fig. 6a–f; text-fig. 123a–b.

? 1920 *Climacograptus Törnquisti* Elles et Wood; Gortani,
    p. 15, pl. 1, figs 16–19.
. 1936 *Climacograptus törnquisti* Elles and Wood;
    Bulman, p. 20; text-fig. 1*a–e*.
  1965 *Rhaphidograptus toernquisti* (Elles & Wood,
    1906); Stein, p. 180, text-figs 16i, 26a–d.
? 1966 *Rhaphidograptus toernquisti* (Elles et Wood), 1906;
    Obut and Sobolevskaya, p. 23, pl. 4, fig. 10;
    text-fig. 14.
  1967 *Rhaphidograptus toernquisti* (Elles et Wood),
    1906; Obut *et al.*, p. 77, pl. 6, fig. 18; pl. 7, fig. 1.
. 1970 *Rhaphidograptus toernquisti* (Elles and Wood,
    1906); Hutt *et al.*, p. 7, pl. 1, figs 21–22.
. 1970 *Rhaphidograptus toernquisti* (Elles and Wood,
    1906); Rickards, p. 54, text-fig. 13, figs 1–3.
. 1974 *Rhaphidograptus toernquisti* (Elles and Wood,
    1906); Hutt, p. 53, pl. 9, figs 1–2; text-fig. 13,
    figs 7–9.
. 1975 *Rhaphidograptus toernquisti* (Elles and Wood,
    1906); Bjerreskov, p. 43, pl. 6, figs C–D.
. 1979 *Rhaphidograptus törnquisti* (Elles & Wood, 1906);
    Jaeger and Robardet, pl. 2, fig. 14.
  1982 *Rhaphidograptus toernquisti* (Elles & Wood);
    Howe, pl. 1, figs g–i.
. 1993 *Rhaphidograptus toernquisti* (Elles and Wood,
    1906); Štorch and Serpagli, p. 30, pl. 5, figs 2, 4;
    text-fig. 8E.
. 1994 *Rhaphidograptus toernquisti* (Elles and Wood);
    Lenz and Vaughan, fig. 3F, H–I.
. 1994 *Rhaphidograptus toernquisti* (Elles and Wood,
    1906); Zalasiewicz and Tunnicliff, p. 711,
    text-fig. 8E–J.
. 1996 *Rhaphidograptus toernquisti* (Elles and Wood,
    1906); Koren' and Rickards, p. 93, pl. 13, figs 4–5;
    text-figs 21D–E, 22L, N–P.
v. 1998 *Rhaphidograptus toernquisti* (Elles & Wood);
    Loydell *et al.*, fig. 6a.
. 1998 *Rhaphidograptus toernquisti* (Elles and Wood,
    1906); Melchin, text-fig. 3FF–GG.
. 1999 *Rhaphidograptus toernquisti* (Elles & Wood, 1906);
    Gutiérrez-Marco *et al.*, pl. 1, fig. 16.
v. 2003 *Rhaphidograptus toernquisti* (Elles & Wood);
    Loydell *et al.*, fig. 4a.
. 2003 *Rhaphidograptus toernquisti* (Elles and Wood);
    Masiak *et al.*, figs 5o, 8k.

Lectotype. Selected by Přibyl 1948, p. 20; figured by Elles and Wood (1906, pl. 26, fig. 1*f*), from the Birkhill Shales of Dob's Linn. Moffat, southern Scotland.

Material. Two proximal ends, from Locality 7.

Description. The proximal region is not well preserved. The sicula is 1.7 mm long in one specimen; its length cannot be measured in the other. There is a single uniserial theca; DVW at the level of th1^1 is 0.5 mm. At th5^1 DVW is 1.25–1.4 mm, 2TRD at th2^1 is 2.0 mm; at th5^1 it is 1.7–1.9 mm. Thecae are of climacograptid type with supragenicular walls very gently inclined to the

rhabdosome axis. Preservation is not good enough to determine details of the median septum.

Remarks. The specimens match previous descriptions of this well-known species. Bjerreskov (1975) noted that the virgella is of spiral form. Bjerreskov (1976) described 'reverse' synrhabdosomes of *R. toernquisti*, in which the rhabdosomes are attached by their virgellae. Zalasiewicz and Tunnicliff (1994) described abnormal virgellae and antivergellar sicular outgrowths in Welsh specimens.

Stratigraphical range. R. toernquisti is a long-ranging species, known from the base of the *vesiculosus* Biozone through to the lower *sedgwickii* Biozone (Rickards 1970). Hutt (1974) noted that the level of appearance of the median septum on the reverse side of the rhabdosome is of some stratigraphical significance (as had been suggested previously, but not quantified, by Rickards 1970): at the fourth to sixth thecal pair in specimens from the upper *vesiculosus* and *cyphus* biozones; at the ninth thecal pair in specimens from the *triangulatus* and *magnus* biozones; and at the ninth to eleventh thecal pair in specimens from the *argenteus* (= *leptotheca*) Biozone.

Genus DIMORPHOGRAPTUS Lapworth, 1876

Dimorphograptus confertus (Nicholson, 1868*a*)
Text-figure 21K

*. 1868*a* *Diplograpsus confertus*, Nich.; Nicholson, p. 526,
    pl. 19, figs 14–15.
. 1908 *Dimorphograptus confertus* (Nicholson); Elles and
    Wood, p. 349, pl. 35, fig. 3*a–d*; text-fig. 227*a–b*.
  1965 *Dimorphograptus confertus* (Nicholson), 1868; Obut
    and Sobolevskaya, p. 21, pl. 4, fig. 9; text-fig. 13.
. 1970 *Dimorphograptus confertus confertus* (Nicholson,
    1868); Rickards, p. 50, pl. 3, fig. 11.
. 1974 *Dimorphograptus confertus confertus* (Nicholson,
    1868); Hutt, p. 50, pl. 8, figs 4, 7–8; pl. 9, fig. 7;
    text-fig. 13, fig. 3.
. 1975 *Dimorphograptus confertus confertus* (Nicholson,
    1869); Bjerreskov, p. 41, pl. 6, fig. A; text-fig.
    13B–C.
. 1986 *Dimorphograptus confertus* (Nicholson); Štorch,
    pl. 6, fig. 3.
non 1990 *Bulmanograptus confertus* (Nicholson); Fang
    et al., p. 98, pl. 25, fig. 13; pl. 6, figs 4–5, 14;
    pl. 27, fig. 5.
. 1994 *Dim. confertus* (Nicholson); Štorch, fig. 5T.
  1999 *Dimorphograptus confertus confertus* (Nicholson,
    1868); Gutiérrez-Marco *et al.*, pl. 1, fig. 9.
v. 2003 *Dimorphograptus confertus* (Nicholson); Loydell
    et al., fig. 3m.
  2003 *Dimorphograptus confertus* (Nicholson); Masiak
    et al., figs 6i, 8i.
v. 2005 *Dimorphograptus confertus confertus* (Nicholson,
    1868); Lüning *et al.*, fig. 15c.

Lectotype. Selected by Bjerreskov (1975, p. 42), the specimen figured by Nicholson (1868a, pl. 19, figs 14–15), from the Coniston Flags of Skelgill Beck, Lake District, northern England. Strachan (1997, p. 83) suggested that this may be NHM H3026.

Material. One flattened specimen from a depth of 30.0 m, core BG-14; *vesiculosus* Biozone.

Description. The proximal end is missing. The preserved uniserial portion bears two thecae. DVW at the second uniserial theca is 0.8 mm; 2TRD is 1.65 mm. The biserial portion increases in DVW from 2.2 to 2.8 mm over six thecal pairs; 2TRD varies between 1.25 and 1.5 mm. The thecae are simple tubes, with concave ventral walls proximally. They increase in inclination from 25 degrees in the uniserial portion to *c.* 50 degrees at the distal end of the biserial portion. The thecal apertural margins undulate slightly and are denticulate.

Remarks. Dimorphograptus confertus is the only species of the genus recorded from Jordan and from Bohemia (Štorch 1994). By contrast, Strachan (1997) listed ten species from the British Isles. *D. swanstoni* Lapworth, 1876 was suggested by Hutt (1974, p. 51) to be possibly conspecific with *D. confertus. D. swanstoni* differs from *D. confertus*, however, not only in having a longer uniserial portion (five or six, rather than three or four thecae), but also in having more widely spaced thecae (see, e.g. Elles and Wood 1908, pl. 35; Lenz 1982, fig. 3G–J). Fang *et al.*'s (1990) specimens are rejected as *D. confertus* on the basis of the uniserial portion being too long.

Stratigraphical range. Štorch (1994, fig. 2) recorded *D. confertus* from all but the lowermost part of the *vesiculosus* Biozone of Bohemia. Bjerreskov (1975, fig. 7) indicated a range throughout her *acinaces* Biozone (equivalent here to the *vesiculosus* Biozone), although it is possible that the lowermost part of the biozone is not exposed on Bornholm (Bjerreskov 1975, p. 7). In Britain, a longer range has been reported, from the *atavus* (= lower *vesiculosus*) Biozone to *cyphus* Biozone (Rickards 1970; Hutt 1974), although in the Lake District *D. confertus* does not appear until the upper *atavus* Biozone (Hutt 1974, p. 50) in which it comprises *c.* 10 per cent of the assemblage, being less common at higher levels.

Genus ATAVOGRAPTUS Rickards, 1974

Atavograptus atavus (Jones, 1909)
Text-figure 21A, H

*.   1909 *Monograptus atavus*, sp. nov. Jones, p. 531,
              text-fig. 18a–d.
? 1966 *Přibylograptus atavus* (Jones), 1909; Obut and
              Sobolevskaya, p. 33, pl. 6, fig. 8; text-fig. 24.
      1970 *Monograptus atavus* Jones; Churkin and Carter,
              p. 36, pl. 3, fig. 11, text-fig. 14E.
.   1970 *M. atavus* Jones; Hutt and Rickards, fig. 3a–b.

.   1974 *Atavograptus atavus* (Jones); Rickards, pl. 9, figs 1–2.
.   1975 *Monograptus atavus* Jones, 1909; Bjerreskov, p. 44,
              pl. 6, figs G–H [with further synonymy].
.   1975 *Atavograptus atavus* (Jones, 1909); Hutt, p. 62,
              pl. 11, fig. 1; pl. 12, figs 5, 9–10.
      1978 *Pristiograptus atavus* (Jones); Chen and Lin, p. 53,
              pl. 9, fig. 27.
      1986 *Atavograptus atavus* (Jones); Štorch, pl. 6, figs 1, 4.
      1988 *Atavograptus atavus* (Jones, 1909); Štorch, p. 12,
              pl. 1, figs 1–3; text-fig. 2A.
      1989 *Atavograptus atavus* (Jones); Melchin, fig. 6A–B.
      1990 *Pristiograptus atavus* (Jones); Fang *et al.*, p. 104,
              pl. 28, fig. 9.
      1991 *Atavograptus atavus*; Gutiérrez-Marco and
              Robardet, fig. 2k–l.
.   1994 *Atavograptus atavus* (Jones); Zalasiewicz and
              Tunnicliff, text-fig. 10A–B.
      1996 *Atavograptus atavus* (Jones); Rickards *et al.*,
              p. 114, figs 6a–c, 11B.
.   1997 *Atavograptus atavus* (Jones, 1909); Koren' and
              Bjerreskov, p. 10, figs 5A–B, 6, 7A–D, F [with
              further synonymy].
.   1997 *Atavograptus atavus* (Jones, 1909); Lukasik and
              Melchin, p. 1139, figs 4A, 8A.
v. 2003 *Atavograptus atavus* (Jones); Loydell *et al.*, figs 3j, 4b.
      2003 *Atavograptus atavus* (Jones); Masiak *et al.*, figs
              4a–b, 8e.

Lectotype. Designated by Přibyl and Spasov (1955, p. 195): BGS GSM 23710, figured by Jones (1909, text-fig. 18b), from the Llandovery of the Rheidol Gorge, Wales.

Material. 14 fragmentary specimens from depths of 27.5 and 30.0 m, core BG-14, and from Locality 7.

Description. Rhabdosome fragments are straight or dorsally curved. No proximal ends are present. Thecae are simple tubes, with a slight geniculum and generally gently convex supragenicular walls. Maximum DVW is 1.15 mm, 2TRDs are 2.1–2.5 mm.

Remarks. The Jordanian material matches previous descriptions of this well-known and widespread species. Isolated material of *At. atavus* was described by Lukasik and Melchin (1997).

Stratigraphical range. At. atavus is a long-ranging species known from the *vesiculosus* Biozone up to the *magnus* Biozone (Hutt 1975).

Genus HUTTAGRAPTUS Koren' and Bjerreskov, 1997

Huttagraptus acinaces (Törnquist, 1899)
Text-figures 21I, 22

*.   1899 *Monograptus acinaces* n. sp. Törnquist, p. 5, pl. 1,
              figs 7–8.

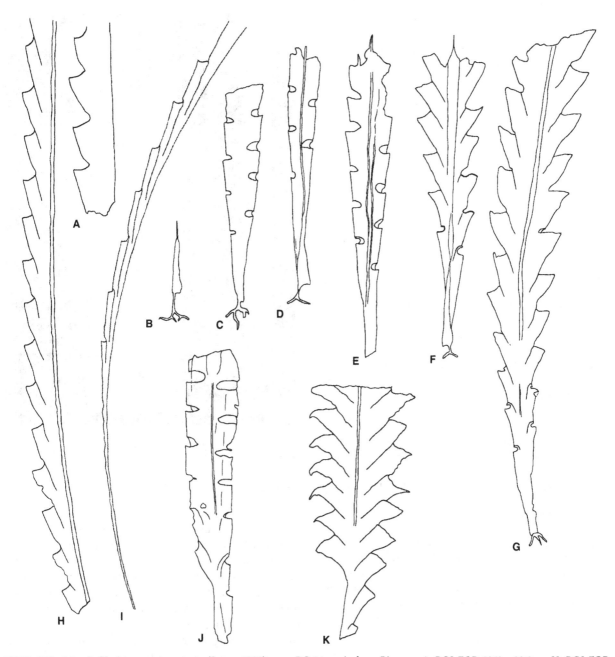

TEXT-FIG. 21. A, H, *Atavograptus atavus* (Jones, 1909), core BG-14, *vesiculosus* Biozone. A, BGS FOR 5369a, 30.0 m. H, BGS FOR 5364, 27.5 m. B–E, *Akidograptus ascensus* Davies, 1929, core WS-6, middle *ascensus-acuminatus* Biozone. B, BGS FOR 5469, 1399.45 m. C, BGS FOR 5464f, 1398.3 m. D, BGS FOR 5470a, 1399.45 m. E, BGS FOR 5470b, 1399.45 m. F–G, *Parakidograptus acuminatus* (Nicholson, 1867), core BG-14, 36.0 m, upper *ascensus-acuminatus* Biozone. F, BGS FOR 5360a, specimen figured by Lüning *et al.* (2005, fig. 15a). G, BGS FOR 5360b. I, *Huttagraptus acinaces* (Törnquist, 1899), BGS FOR 5463c, Locality 10, *vesiculosus* Biozone. J, *Rhaphidograptus toernquisti* (Elles and Wood, 1906), GD-NRA 18401-54g, Locality 7, *vesiculosus* Biozone. K, *Dimorphograptus confertus* (Nicholson, 1868a), BGS FOR 5361a, core BG-14, 30.0 m, *vesiculosus* Biozone. All figures × 10.

. 1909 *Monograptus rheidolensis*, sp. nov. Jones, p. 535, text-fig. 19*a–d*.

. 1911 *Monograptus acinaces*, Törnquist; Elles and Wood, p. 364, pl. 36, figs 2*a–d*; text-fig. 237*a–d*.

1970 *Monograptus acinaces* Törnquist; Churkin and Carter, p. 35, pl. 3, figs 13–15; text-fig. 14*A*.

. 1970 *M. acinaces* Törnquist; Hutt and Rickards, fig. 3i–j.

. 1975 *Lagarograptus acinaces* (Törnquist, 1899); Hutt, p. 69, pl. 13, figs 5–6; text-fig. 16, figs 1–3.

? 1982 *Lagarograptus acinaces* (Törnquist); Lenz and McCracken, fig. 5h.

1989 *Lagarograptus acinaces* (Törnquist); Melchin,
   fig. 7F.
. 1997 *Huttagraptus acinaces* (Törnquist, 1899); Koren'
   and Bjerreskov, p. 18, figs 5D–F, 11–13.
v 1998 *Huttagraptus acinaces* (Törnquist); Loydell *et al.*,
   fig. 6b.
? 1998 *Lagarograptus acinaces* (Törnquist); Underwood
   et al., fig. 5JJ–MM.
v 2003 *Huttagraptus acinaces* (Törnquist); Loydell *et al.*,
   fig. 3i.

Holotype. By monotypy: LO 1436t, figured by Törnquist (1899, pl. 1, figs 7–8), Hutt and Rickards (1970, fig. 3i), and Koren' and Bjerreskov (1997, figs 11T, 12H); from the Llandovery of Röstånga, Sweden.

Material. Two specimens: one long specimen from Locality 10; one short fragment from a depth of 30.0 m, core BG-14. There are several other short *Huttagraptus* fragments in the 30.0 m core sample. Palynological processing of a subsample from 30.0 m yielded two isolated fragments (one is illustrated in Text-fig. 22).

Description. The long rhabdosome fragment is 20 mm long, but is longitudinally damaged proximally for more than half of this length. The rhabdosome exhibits broad dorsal curvature. It increases in width from 0.4 to 0.6 mm over the seven preserved thecae. Over the same distance 2TRD decreases from 2.7 to 2.6 mm. The short fragment is almost straight, has a DVW of 0.8 mm and a 2TRD of 2.5–2.7 mm. In both specimens the thecae are long tubes, overlapping for more than half their length. There are conspicuous apertural excavations on the thecae of the longer fragment and these thecae have straight to very slightly concave supragenicular walls. The more distal thecae are more pristiograptid in form.

The longer of the three isolated fragments is *c.* 8 mm long, with a DVW of 0.65 mm; 2TRD is 2.05–2.65 mm. The shorter is a distal fragment, with a DVW of 1.1 mm and a 1TRD of 1.25 mm. In both fragments thecae are of pristiograptid form. Interthecal septa are visible in the shorter fragment and indicate thecal overlap of about one-half. Thecal apertures are horizontal.

Remarks. This species was described and discussed in considerable detail by Koren' and Bjerreskov (1997).

Stratigraphical range. *H. acinaces* is known from the *vesiculosus* and *cyphus* biozones (Hutt 1975). In the Billegrav-1 core, Bornholm, Koren' and Bjerreskov (1997) recorded the first appearance datum (FAD) of *H. acinaces* low in the *vesiculosus* Biozone below that of *Atavograptus atavus*.

CONCLUSIONS

The following are some of the main conclusions drawn from this work.

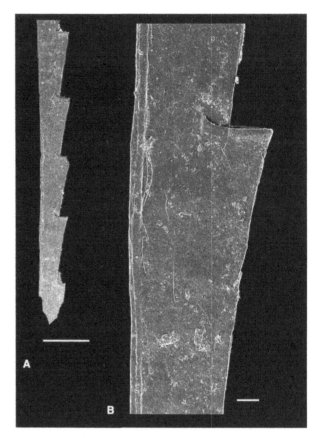

TEXT-FIG. 22. *Huttagraptus acinaces* (Törnquist, 1899), BGS FOR 5473; core BG-14, 30.0 m, *vesiculosus* Biozone. A, rhabdosome fragment of five thecae. B, close-up of theca; note slightly concave ventral wall. Scale bars represent 1 mm in A, 100 μm in B.

1. The upper Hirnantian and lower Rhuddanian of Jordan yield diverse assemblages of graptoloids. Approximately half of the species described have been recorded previously from peri-Gondwanan Europe; the remainder are new, or have been left in open nomenclature, or have been described previously from localities only outside of peri-Gondwanan Europe, e.g. North Africa (*Ne. africanus*, *Ne.* sp. 3, *N. targuii*, *Pa. libycus*), China (*Ne. shanchongensis*) and Uzbekistan (*N. larini*).

2. The presence of *Ne. africanus* in Jordan, in assemblages containing more widespread, biostratigraphically important taxa, enhances correlation of the North African *africanus* and *tariti* Biozone with the 'standard' graptolite biozonation. *Ne. africanus* appears first in the BG-14 core in an unzoned interval only 1 m above a level bearing *P. acuminatus*, suggesting that the base of the *africanus* and *tariti* Biozone may lie close to the *ascensus-acuminatus/vesiculosus* Biozone boundary. Legrand (2003) had tentatively correlated the base of this biozone with the middle *vesiculosus* Biozone. It appears that *Ne.* sp. 3 has a longer stratigraphical range in Jordan than it has in North Africa and is not suitable as a biozonal index species.

3. There appear to be no confirmed occurrences of *Normalograptus persculptus* in the Silurian. The species seems to be confined to the upper Hirnantian *persculptus* Biozone.

4. The stratigraphical range of *N. transgrediens* is longer than previously thought, extending into the *vesiculosus* Biozone; thus, some re-assessment may be required of the biostratigraphy of areas (e.g. the Oslo region, Norway) where the presence of *N. transgrediens* has been taken to indicate the *ascensus-acuminatus* Biozone.

5. Elles and Wood's (1906) misleading description of *Normalograptus normalis* has resulted in considerable confusion surrounding the identification of this species. Many records of *N. normalis* are the narrower species, *N. ajjeri*. The name *N. normalis* should be used only for specimens similar to the type material.

6. *Normalograptus rectangularis* is a useful species biostratigraphically; it appears in the upper *ascensus-acuminatus* Biozone and thus enables recognition of levels significantly above the Ordovician/Silurian boundary.

Acknowledgements. The following very kindly loaned material to me for this study: Sebastian Lüning (University of Bremen), Barrie Rickards (Sedgwick Museum, Cambridge), Howard Armstrong (University of Durham), Claire Mellish and Adrian Rushton (Natural History Museum, London). Anthony Butcher (University of Portsmouth) processed the sample that yielded the *H. acinaces* of Text-figure 22. Petr Štorch generously provided comparative material of Bohemian *Neodiplograptus* taxa and digital images of the holotype of *Normalograptus angustus*. This paper benefited from the constructive reviews of Mike Melchin and Petr Štorch, and the editorial advice of Phil Lane.

REFERENCES

APOLLONOV, M. K., BANDALETOV, S. M. and NIKITIN, J. F. 1980. *The Ordovician-Silurian boundary in Kazakhstan.* Nauka Kazakhstan SSR, Alma Ata, 232 pp., 56 pls. [In Russian].

—— KOREN', T. N., NIKITIN, J. F., PALETZ, L. M. and TZAI, D. T. 1988. Nature of the Ordovician-Silurian boundary in south Kazakhstan, USSR. *Bulletin of the British Museum (Natural History), Geology Series*, **43**, 133–138.

ARMSTRONG, H. A., TURNER, B. R., MAKHLOUF, I. M., WEEDON, G. P., WILLIAMS, M., AL SMADI, A. and ABU SALAH, A. 2005. Origin, sequence stratigraphy and depositional environment of an upper Ordovician (Hirnantian) deglacial black shale, Jordan. *Palaeogeography, Palaeoclimatology, Palaeoecology*, **220**, 273–289.

—— —— —— —— —— —— —— 2006. Reply to 'Origin, sequence stratigraphy and depositional environment of an upper Ordovician (Hirnantian) deglacial black shale, Jordan'. *Palaeogeography, Palaeoclimatology, Palaeoecology*, **230**, 356–360.

ARMSTRONG, J., YOUNG, J. and ROBERTSON, D. 1876. *Catalogue of the western Scottish fossils.* Blackie & Son, Glasgow, 164 pp., 4 pls.

BAILLIE, P. W., BANKS, M. R. and RICKARDS, R. B. 1978. Early Silurian graptolites from Tasmania and their significance. *Search*, **9**, 46–47.

BATES, D. E. B. and KIRK, N. H. 1984. Autecology of Silurian graptolites. 121–139. *In* BASSETT, M. G. and LAWSON, J. D. (eds). Autecology of Silurian organisms. *Special Papers in Palaeontology*, **32**, 1–295.

BENDER, F. 1963. Stratigraphie der 'Nubischen Sandsteine' in Süd-Jordanien. *Geologisches Jahrbuch*, **81**, 237–275.

BENTON, M. J. 1979. *H. A. Nicholson (1844–99), invertebrate palaeontologist: bibliography and catalogue of his type and figured material.* Royal Scottish Museum Information Series, Geology, **7**. Edinburgh, 94 pp.

BJERRESKOV, M. 1975. Llandoverian and Wenlockian graptolites from Bornholm. *Fossils and Strata*, **8**, 1–94, pls 1–13.

—— 1976. A new type of graptolite synrhabdosome. *Bulletin of the Geological Society of Denmark*, **25**, 41–47.

BULMAN, O. M. B. 1936. *Rhaphidograptus*, a new graptolite genus. *Geological Magazine*, **73**, 19–26.

—— and RICKARDS, R. B. 1968. Some new diplograptids from the Llandovery of Britain and Scandinavia. *Palaeontology*, **11**, 1–15.

CHALETSKAYA, O. N. 1960. New species of the Llandovery graptolites of Central Asia. 373–375. *In* MARKOVSKY, B. P. (ed.). *New species of ancient plants and invertebrates in the USSR. II.* Gosudarstvennoe Nauchno-Teknicheskoye Izdatel'stvo, Moscow. [In Russian].

CHEN XU, FAN JUN-XUAN, MELCHIN, M. J. and MITCHELL, C. E. 2005a. Hirnantian (latest Ordovician) graptolites from the upper Yangtze region, China. *Palaeontology*, **48**, 235–280.

—— and LIN YAO-KUN 1978. Lower Silurian graptolites from Tongzi, northern Guizhou. *Memoirs of the Nanjing Institute of Geology and Palaeontology, Academia Sinica*, **12**, 1–76, pls 1–19. [In Chinese, English abstract].

—— MELCHIN, M. J., SHEETS, H. D., MITCHELL, C. E. and FAN JUN-XUAN 2005b. Patterns and processes of latest Ordovician graptolite extinction and recovery based on data from South China. *Journal of Paleontology*, **79**, 842–861.

—— and QIAN ZE-SHU 1990. Isolated Llandovery graptolites from northern Jiangsu. *Acta Palaeontologica Sinica*, **29**, 1–11, pls 1–4.

—— RONG JIA-YU and FAN JUN-XUAN 2003. A proposal for a candidate section for restudy of the base of Silurian. 119–123. *In* ORTEGA, G. and ACEÑOLAZA, G. F. (eds). *Proceedings of the 7th International Graptolite Conference & Field Meeting of the International Subcommission on Silurian Stratigraphy.* Instituto Superior de Correlación Geológica (INSUGEO), Tucumán, **18**, 181 pp.

CHURKIN, M. Jr and CARTER, C. 1970. Early Silurian graptolites from southeastern Alaska and their correlation with graptolitic sequences in North America and the Arctic. *United States Geological Survey, Professional Paper*, **653**, 1–51, pls 1–4.

CUERDA, A. J., RICKARDS, R. B. and CINGOLANI, C. 1988. A new Ordovician-Silurian boundary section in San Juan Province, Argentina, and its definitive graptolite fauna. *Journal of the Geological Society, London*, **145**, 749–757.

DAVIES, K. A. 1929. Notes on the graptolite faunas of the Upper Ordovician and Lower Silurian. *Geological Magazine*, **66**, 1–27.

DESIO, A. 1940. Fossili neosilurici del Fezzan occidentale. *Annali del Museo Libico di Storia Naturale*, **2**, 13–45, pls 1–3.

ELLES, G. L. and WOOD, E. M. R. 1906. A monograph of British graptolites. Part 5. *Monograph of the Palaeontographical Society*, **60** (288), lxxiii–xcvi, 181–216, pls 26–27.

—— —— 1907. A monograph of British graptolites. Part 6. *Monograph of the Palaeontographical Society*, **61** (297), xcvii–cxx, 217–272, pls 28–31.

—— —— 1908. A monograph of British graptolites. Part 7. *Monograph of the Palaeontographical Society*, **62** (305), cxxi–cxlviii, 273–358, pls 32–35.

—— —— 1911. A monograph of British graptolites. Part 8. *Monograph of the Palaeontographical Society*, **64** (316), 359–414, pls 36–41.

FANG YI-TING, LIANG SHI-JING, ZHANG DA-LIANG and YU JIN-LONG 1990. *Stratigraphy and graptolite fauna of Lishuwo Formation from Wuning, Jiangxi*. Nanjing University Publishing House, Nanjing, 155 pp., 29 pls. [In Chinese, English summary].

GE MEI-YU 1984. The graptolite fauna of the Ordovician–Silurian section in Yuqian, Zhejiang. 389–444, pls 1–9. *In* NANJING INSTITUTE OF GEOLOGY AND PALAEONTOLOGY, ACADEMIA SINICA (ed.). *Stratigraphy and palaeontology of systemic boundaries in China, Ordovician-Silurian boundary (1)*. Anhui Science and Technology Publishing House, Hefei, 516 pp.

GNOLI, M, KŘÍŽ, J., LEONE, F., OLIVIERI, R., SERPAGLI, E. and ŠTORCH, P. 1990. Lithostratigraphic units and biostratigraphy of the Silurian and early Devonian of southwest Sardinia. *Bolletino della Società Paleontologica Italiana*, **29**, 11–23.

GORTANI, M. 1920. Contribuzioni allo studío del Paleozoico carnico. *Palaeontographia Italica*, **26**, 1–56, pls 1–3.

GUTIÉRREZ-MARCO, J.-C. and ROBARDET, M. 1991. Découverte de la zone à *Parakidograptus acuminatus* (base du Llandovery) dans le Silurien du Synclinorium de Truchas (Zone asturo-léonaise, Nord-Ouest de l'Espagne): conséquences stratigraphiques et paléogéographiques au passage Ordovicien–Silurien. *Comptes Rendus de l'Académie des Sciences, Paris, Série 2*, **312**, 729–734.

—— ROQUÉ BERNAL, J., ROBARDET, M. and IBÁÑEZ SOTILLOS, R. 1999. Graptolitos de la Biozona de *Coronograptus cyphus* (Rhuddaniense: Silúrico inferior) en el área del Montseny (Cadenas Costeras Catalanas, noreste de España. *Temas Geológico-Mineros ITGE*, **26**, 618–622.

HISINGER, W. 1837. *Lethaea Suecica seu Petrifacta Sueciae, Supplementum 1*. Holmiae, Stockholm, 124 pp., 2 pls.

HOPKINSON, J. 1869. On British graptolites. *Journal of the Quekett Microscopical Club*, **1**, 151–166, pl. 8.

HOWE, M. P. A. 1982. The lower Silurian graptolites of the Oslo region. 21–32. *In* WORSLEY, D. (ed.). *International Union of Geological Sciences, Subcommission on Silurian Stratigraphy, Field Meeting Oslo region 1982*. Palaeontological Contributions from the University of Oslo, **278**, 1–175.

—— 1983. Measurement of thecal spacing in graptolites. *Geological Magazine*, **120**, 635–638.

HUANG ZHI-GAO 1982. Latest Ordovician and earliest Silurian graptolite assemblages of Xainza district, Xizang (Tibet) and Ordovician-Silurian boundary. 27–52, pls 1–3. *In* EDITORIAL COMMITTEE OF MINISTRY OF GEOLOGY AND MINERAL RESOURCES (ed.). Contribution to the geology of the Qinghai-Xizang (Tibet) plateau, **7**. Geological Publishing House, Beijing, 168 pp. [In Chinese, English abstract].

HUNDT, R. 1942. Beiträge zur Kenntnis des Mitteldeutschen Graptolithenmeeres. *Beiträge zur Geologie von Thüringen*, **6**, 205–231, pls 1–7.

HUTT, J. E. 1974. The Llandovery graptolites of the English Lake District. Part 1. *Monograph of the Palaeontographical Society*, **128** (540), 1–56, pls 1–10.

—— 1975. The Llandovery graptolites of the English Lake District. Part 2. *Monograph of the Palaeontographical Society*, **129** (542), 57–137, pls 11–26.

—— and RICKARDS, R. B. 1970. The evolution of the earliest Llandovery monograptids. *Geological Magazine*, **107**, 67–77.

—— —— and SKEVINGTON, D. 1970. Isolated Silurian graptolites from the Bollerup and Klubbudden stages of Dalarna, Sweden. *Geologica et Palaeontologica*, **4**, 1–23.

JAEGER, H. 1976. Das Silur und Unterdevon vom thüringischen Typ in Sardinien und seine regionalgeologische Bedeutung. *Nova Acta Leopoldina, Neue Folge*, **45**, 263–299.

—— and ROBARDET, M. 1979. Le Silurien et le Dévonien basal dans le Nord de la Province de Seville (Espagne). *Geobios*, **12**, 687–714.

JONES, O. T. 1909. The Hartfell–Valentian succession in the district around Plynlimon and Pont Erwyd (North Cardiganshire). *Quarterly Journal of the Geological Society of London*, **65**, 463–537, pls 24–25.

JONES, W. D. V. and RICKARDS, R. B. 1967. *Diplograptus penna* Hopkinson 1869, and its bearing on vesicular structures. *Paläontologische Zeitschrift*, **41**, 173–185.

KOREN', T. N., AHLBERG, P. and NIELSEN, A. T. 2003. The post-*persculptus* and pre-*ascensus* graptolite fauna in Scania, south-western Sweden: Ordovician or Silurian? 133–138. *In* ORTEGA, G. and ACEÑOLAZA, G. F. (eds). *Proceedings of the 7th International Graptolite Conference & Field Meeting of the International Subcommission on Silurian Stratigraphy*. Instituto Superior de Correlación. Geológica (INSUGEO), Tucumán, Serie Correlación Geológica, **18**, 181 pp.

—— and BJERRESKOV, M. 1997. Early Llandovery monograptids from Bornholm and the southern Urals: taxonomy and evolution. *Bulletin of the Geological Society of Denmark*, **44**, 1–43.

—— and MELCHIN, M. J. 2000. Lowermost Silurian graptolites from the Kurama Range, eastern Uzbekistan. *Journal of Paleontology*, **74**, 1093–1113.

—— MIKHAYLOVA, N. F. and TZAI, D. T. 1980. Class Graptolithina. Graptolites. 121–170, pls 33–54. *In* APOLLO-

NOV, M. K., NIKITIN, I. F. and BANDALETOV, S. M. (eds). *The Ordovician-Silurian boundary in Kazakhstan.* 'Nauka' Kazakh SSR, Alma-Ata, 300 pp., 56 pls. [In Russian].

—— —— ORADOVSKAYA, M. M., PYLMA, L. J., SOBOLEVSKYA, R. F. and CHUGAEVA, M. N. 1983. *The Ordovician and Silurian boundary in the northeast of the USSR.* 'Nauka', Leningrad, 208 pp., 48 pls. [In Russian].

—— —— and SOBOLEVSKAYA, R. F. 1988. The Ordovician-Silurian boundary beds of the north-east USSR. *Bulletin of the British Museum (Natural History), Geology Series*, **43**, 133–138.

—— and RICKARDS, R. B. 1996. Taxonomy and evolution of Llandovery biserial graptoloids from the southern Urals, western Kazakhstan. *Special Papers in Palaeontology*, **54**, 1–103.

—— —— 2004. An unusually diverse Llandovery (Silurian) diplograptid fauna from the southern Urals of Russia and its evolutionary significance. *Palaeontology*, **47**, 859–918.

LAPWORTH, C. 1873. Notes on the British graptolites and their allies. 1. On an improved classification of the Rhabdophora. *Geological Magazine, 1*, **10**, 500–504, 555–560, table 1.

—— 1876. On Scottish Monograptidae. *Geological Magazine, 2*, **3**, 308–321, 350–360, 499–507, 544–552, pls 10–13, 20.

—— 1877. On the graptolites of County Down. *Proceedings of the Belfast Naturalists' Field Club*, **1876–1877**, 125–148, pls 5–7.

LAPWORTH, H. 1900. The Silurian sequence of Rhayader. *Quarterly Journal of the Geological Society of London*, **56**, 67–137, pls 6–7.

LEGRAND, P. 1970. Les couches à *Diplograptus* du Tassili de Tarit (Ahnet, Sahara algérien). *Bulletin de la Société d'Histoire Naturelle de l'Afrique du Nord*, **60**, 3–58.

—— 1977. Contribution à l'etude des graptolites du Llandoverien inférieur de l'Oued In Djerane (Tassili N'ajjer oriental, Sahara algérien). *Bulletin de la Société d'Histoire Naturelle de l'Afrique du Nord*, **67**, 141–196.

—— 1986. The lower Silurian graptolites of Oued In Djerane: a study of populations at the Ordovician-Silurian boundary. 145–153. *In* HUGHES, C. P. and RICKARDS, R. B. (eds). *Palaeoecology and biostratigraphy of graptolites.* Geological Society, Special Publication, **20**, 277 pp.

—— 1987. Modo de desarrollo del Suborden Diplograptina (Graptolithina) en el Ordovícico Superior y en el Silúrico. Implicaciones taxonómicas. *Revista Española de Paleontología*, **2**, 59–64.

—— 1993. Graptolites d'âge ashgillien dans la région de Chirfa (Djado, République du Niger). *Bulletin des Centres de Recherches Exploration-Production Elf Aquitaine*, **17**, 435–442.

—— 1995. Á propos d'un niveau à *Neodiplograptus* dans le Silurien inférieur à l'est de Ouallene, Asejrad (Sahara algérien), implications stratigraphiques et paléogéographiques. *118 Congrès national des Sociétés historiques et scientifiques, 4ème colloque sur la Géologie africaine, Pau*, 409–424.

—— 1999. Approche stratigraphique de l'Ordovicien terminal et du Silurien inferieur du Sahara algerien par l'etude des Diplograptides (Graptolites). Unpublished PhD thesis, Université Michel de Montaigne–Bordeaux III, Institut *EGID* Bordeaux III.

—— 2000. Une région de référence pour la limite Ordovicien-Silurien: l'Oued In Djerane, Sahara algérien. *Comptes Rendus de l'Académie des Sciences, Paris, Sciences de la Terre et des Planètes*, **330**, 61–66.

—— 2001. La faune graptolitique de la région d'In Azaoua (Tassili Oua-n-Ahaggar, confines algéro-nigériens). *Annales de la Société Géologique du Nord*, **8**, 137–158, pl. 12.

—— 2002. La Formation des Argiles de Tedjert (?Ordovicien terminal–Silurien inferieur) au Tassili Oua-n-Ahaggar oriental (Sahara algerien) et sa faune graptolitique. *Annales de la Société Géologique du Nord, Série 2*, **9**, 215–229, pl. 12.

—— 2003. Silurian stratigraphy and paleogeography of the northern African margin of Gondwana. *New York State Museum Bulletin*, **493**, 59–104.

LENZ, A. C. 1982. Llandoverian graptolites of the northern Canadian Cordillera: *Petalograptus, Cephalograptus, Rhaphidograptus, Dimorphograptus*, Retiolitidae, and Monograptidae. *Life Sciences Contributions, Royal Ontario Museum*, **130**, 1–154.

—— and McCRACKEN, A. D. 1982. The Ordovician-Silurian boundary, northern Canadian Cordillera: graptolite and conodont correlation. *Canadian Journal of Earth Sciences*, **19**, 1308–1322.

—— and VAUGHAN, A. P. M. 1994. A Late Ordovician to middle Wenlockian graptolite sequence from a borehole within the Rathkenny Tract, eastern Ireland, and its relation to the paleogeography of the Iapetus Ocean. *Canadian Journal of Earth Sciences*, **31**, 608–616.

LI JI-JIN 1984. Graptolites across the Ordovician-Silurian boundary from Jingxian, Anhui. 309–370, pls 1–18. *In* NANJING INSTITUTE OF GEOLOGY AND PALAEONTOLOGY, ACADEMIA SINICA (ed.). *Stratigraphy and palaeontology of systemic boundaries in China, Ordovician-Silurian boundary (1).* Anhui Science and Technology Publishing House, Hefei, 516 pp.

—— 1990. Discovery of monograptids in basal part of lower Silurian from S. Anhui with special reference to their origin. *Acta Palaeontologica Sinica*, **29**, 204–215, pl. 1.

—— and GE MEI-YU 1981. Development and systematic position of akidograptids. *Acta Palaeontologica Sinica*, **20**, 225–234, pl. 1.

LIN YAO-KUN and CHEN XU 1984. *Glyptograptus persculptus* Zone – the earliest Silurian graptolite zone from Yangzi Gorges, China. 203–225, pls 1–6. *In* NANJING INSTITUTE OF GEOLOGY AND PALAEONTOLOGY, ACADEMIA SINICA (ed.). *Stratigraphy and palaeontology of systemic boundaries in China, Ordovician-Silurian boundary (1).* Anhui Science and Technology Publishing House, Hefei, 516 pp.

LOYDELL, D. K. 1991. Isolated graptolites from the Llandovery of Kallholen (Sweden). *Palaeontology*, **34**, 671–693.

—— KALJO, D. and MÄNNIK, P. 1998. Integrated biostratigraphy of the lower Silurian of the Ohesaare core, Saaremaa, Estonia. *Geological Magazine*, **135**, 769–783.

—— MALLETT, A., MIKULIC, D. G., KLUESSENDORF, J. and NORBY, R. D. 2002. Graptolites from near the Ordovician-Silurian boundary in Illinois and Iowa. *Journal of Paleontology*, **76**, 134–137.

—— MÄNNIK, P. and NESTOR, V. 2003. Integrated biostratigraphy of the lower Silurian of the Aizpute-41 core, Latvia. *Geological Magazine*, **140**, 205–229.

LUKASIK, J. J. and MELCHIN, M. J. 1997. Morphology and classification of some Early Silurian monograptids (Graptoloidea) from the Cape Phillips Formation, Canadian Arctic Islands. *Canadian Journal of Earth Sciences*, **34**, 1128–1149.

LÜNING, S., KOLONIC, S., LOYDELL, D. K. and CRAIG, J. 2003. Reconstruction of the original organic richness in weathered Silurian shale outcrops (Murzuq and Kufra basins, southern Libya). *GeoArabia*, **8**, 299–308.

—— LOYDELL, D., ŠTORCH, P., SHAHIN, Y. M. and CRAIG, J. 2006. Origin, sequence stratigraphy and depositional environment of an upper Ordovician (Hirnantian) deglacial black shale, Jordan – discussion. *Palaeogeography, Palaeoclimatology, Palaeoecology*, **230**, 352–355.

—— SHAHIN, Y. M., LOYDELL, D., AL-RABI, H. T., MASRI, A., TARAWNEH, B. and KOLONIC, S. 2005. Anatomy of a world-class source rock: distribution and depositional model of Silurian organic-rich shales in Jordan and implications for hydrocarbon potential. *American Association of Petroleum Geologists, Bulletin*, **89**, 1397–1427.

MALETZ, J. 1999. Lowermost Silurian graptolites of the Deerlijk 404 well, Brabant Massif (Belgium). *Neues Jahrbuch für Geologie und Paläontologie, Abhandlungen*, **213**, 335–354.

MANCK, E. 1923. Untersilurische Graptolithenarten der Zone 10 des Obersilurs, ferner *Diversograptus* gen. nov. sowie einige neue Arten anderer Gattungen. *Natur (Leipzig)*, **14**, 282–289.

MASIAK, M., PODHALAŃSKA, T. and STEMPIEŃ-SAŁEK, M. 2003. Ordovician-Silurian boundary in the Bardo Syncline, Holy Cross Mountains, Poland – new data on fossil assemblages and sedimentary succession. *Geological Quarterly*, **47**, 311–330.

MAURY, C. J. 1929. Uma zona de Graptolitos do Llandovery inferior no Rio Trombetas, Estado do Pará, Brasil. *Servico Geologico e Mineralogico do Brasil, Monographia*, **7**, 6–45.

McCOY, F. 1850. On some new genera and species of Silurian Radiata in the collection of the University of Cambridge. *Annals and Magazine of Natural History*, 2, **6**, 270–290.

MELCHIN, M. J. 1989. Llandovery graptolite biostratigraphy and paleobiogeography, Cape Phillips Formation, Canadian Arctic Islands. *Canadian Journal of Earth Sciences*, **26**, 1726–1746.

—— 1998. Morphology and phylogeny of some early Silurian 'diplograptid' genera from Cornwallis Island, Arctic Canada. *Palaeontology*, **41**, 263–315.

—— 2003. Restudying a global stratotype for the base of the Silurian: a progress report. 147–149. *In* ORTEGA, G. and ACEÑOLAZA, G. F. (eds). *Proceedings of the 7th International Graptolite Conference & Field Meeting of the International Subcommission on Silurian Stratigraphy*. Instituto Superior de Correlación Geológica (INSUGEO), Tucumán, Serie Correlación Geológica, **18**, 181 pp.

—— HOLMDEN, C. and WILLIAMS, S. H. 2003. Correlation of graptolite biozones, chitinozoan biozones, and carbon isotope curves through the Hirnantian. 101–104. *In*

ALBANESI, G. L., BERESI, M. S. and PERALTA, S. H. (eds). *Ordovician from the Andes*. Instituto Superior de Correlación Geológica (INSUGEO), Tucumán, Serie Correlación Geológica, **17**, 549 pp.

—— McCRACKEN, A. D. and OLIFF, F. J. 1991. The Ordovician-Silurian boundary on Cornwallis and Truro islands, Arctic Canada: preliminary data. *Canadian Journal of Earth Sciences*, **28**, 1854–1862.

—— and MITCHELL, C. E. 1991. Late Ordovician extinction in the Graptoloidea. 143–156. *In* BARNES, C. R. and WILLIAMS, S. H. (eds). *Advances in Ordovician geology*. Geological Survey of Canada, Paper, **90-9**, 335 pp.

—— and WILLIAMS, S. H. 2000. A restudy of the akidographtine graptolites from Dob's Linn and a proposed redefined zonation of the Silurian stratotype. 63. *In* SIMPSON, A. and WINCHESTER-SEETO, T. (eds). *Palaeontology downunder 2000*. Geological Society of Australia, Sydney, 183 pp.

MITCHELL, C. E., MELCHIN, M. J., SHEETS, D. H., CHEN XU and FAN JUN-XUAN 2003. Was the Yangtze Platform a refugium for graptolites during the Hirnantian (Late Ordovician) mass extinction. 523–526. *In* ALBANESI, G. L., BERESI, M. S. and PERALTA, S. H. (eds). *Ordovician from the Andes*. INSUGEO, Serie Correlación Geológica, **17**, 549 pp.

MU EN-ZHI and LIN YAO-KUN 1984. Graptolites from the Ordovician-Silurian boundary sections of Yichang area, W. Hubei. 45–73, pls 1–8. *In* NANJING INSTITUTE OF GEOLOGY AND PALAEONTOLOGY, ACADEMIA SINICA (ed.). *Stratigraphy and palaeontology of systemic boundaries in China, Ordovician-Silurian boundary (1)*. Anhui Science and Technology Publishing House, Hefei, 516 pp.

—— and NI YU-NAN 1983. Uppermost Ordovician and lowermost Silurian graptolites from the Xainza area of Xizang (Tibet) with discussion on the Ordovician-Silurian boundary. *Palaeontologia Cathayana*, **1**, 151–179.

NICHOLSON, H. A. 1867. On some fossils from the Lower Silurian rocks of the South of Scotland. *Geological Magazine*, 1, **4**, 107–113, pl. 7.

—— 1868a. On the graptolites of the Coniston Flags; with notes on the British species of the genus *Graptolites*. *Quarterly Journal of the Geological Society of London*, **24**, 521–545, pls 19–20.

—— 1868b. On the nature and zoological position of the Graptolitidae. *Annals and Magazine of Natural History*, 4, **1**, 55–61, pl. 3.

—— 1869. On some new species of graptolites. *Annals and Magazine of Natural History*, 4, **4**, 231–242, pl. 11.

OBUT, A. M. 1949. *Field atlas of the leading graptolites of the Upper Silurian of the Kirghiz SSR*. Kirgizskii Filial Akademii Nauk SSSR, Geologicheskii Institut, 57 pp. [In Russian].

—— and SOBOLEVSKAYA, R. F. 1966. *Lower Silurian graptolites of Kazakhstan*. Akademiya Nauk SSR, Sibirskoye Otdelenie Institut Geologii i Geofiziki, Ministerstvo Geologii SSSR, Nauchno-Issledovatel'sky Institut Geologii Arktiki, 52 pp., 8 pls. [In Russian].

—— —— and BONDAREV, V. I. 1965. *Silurian graptolites of Taimir*. Akademiya Nauk SSR, Sibirskoye Otdelenie Institut Geologii i Geofiziki, Ministerstvo Geologii SSSR, Nauchno-Issledovatel'sky Institut Geologii Arktiki, 120 pp., 19 pls. [In Russian].

—— —— and MERKUREVA, A. P. 1968. *Llandovery graptolites from borehole cores in the Noril'sk district.* Akademiya Nauk SSR, Sibirskoye Otdelenie Institut Geologii i Geofiziki, Ministerstvo Geologii SSSR, Nauchno-Issledovatel'sky Institut Geologii Arktiki, 136 pp., 35 pls. [In Russian].

—— —— and NIKOLAEV, A. A. 1967. *Graptolites and stratigraphy of the lower Silurian along the margins of the Kolyma massif.* Akademiya Nauk SSR, Sibirskoye Otdelenie Institut Geologii i Geofiziki, Ministerstvo Geologii SSSR, Nauchno-Issledovatel'sky Institut Geologii Arktiki, 164 pp., 20 pls. [In Russian].

PAŠKEVIČIUS, J. 1979. *Biostratigraphy and graptolites of the Lithuanian Silurian.* Mokslas, Vilnius, 267 pp. [In Russian].

PEDERSEN, T. B. 1922. Rastritesskiffern på Bornholm. *Meddelelser fra Dansk Geologisk Forening,* **6** (11), 1–29.

PERNER, J. 1895. *Études sur les Graptolites de Bohême. IIième Partie. Monographie des Graptolites de l'Étage D.* Raimond Gerhard, Prague, 31 pp., pls 4–8.

PIÇARRA, J. M., ŠTORCH, P., GUTIÉRREZ-MARCO, J. C. and OLIVEIRA, J. T. 1995. Characterization of the *Parakidograptus acuminatus* graptolite Biozone in the Silurian of the Barrancos region (Ossa Morena Zone, south Portugal). *Comunicações Instituto Geológico e Mineiro,* **81**, 3–8.

POWELL, J. H., MOH'D, B. K. and MASRI, A. 1994. Late Ordovician–Early Silurian glaciofluvial deposits preserved in palaeovalleys in South Jordan. *Sedimentary Geology,* **89**, 303–314.

PŘIBYL, A. 1947. Classification of the genus *Climacograptus* Hall, 1865. *Bulletin International de l'Académie Tchèque des Sciences,* **48** (2), 1–12, pls 1–2.

—— 1948. Bibliographic index of Bohemian Silurian graptolites. *Knihovna Státního Geologického Ústavu Československé Republiky,* **22**, 1–96, tables 1–2.

—— 1951. Revision of the Diplograptidae and Glossograptidae of the Ordovician of Bohemia. *Bulletin International de l'Académie Tchèque des Sciences,* **50**, 1–51, pls 1–5.

—— and SPASOV, C. 1955. Bibliographical index of Bulgarian Silurian graptolites. *Izvestiya na Geologiya Instituta, Bulgarii Akademiya Nauk, Sofia,* **3**, 167–209.

RICKARDS, R. B. 1970. The Llandovery (Silurian graptolites) of the Howgill Fells, northern England. *Monograph of the Palaeontographical Society,* **123** (524), 1–108, pls 1–8.

—— 1973. *Climacograptus scalaris* (Hisinger) and the subgenus *Glyptograptus* (*Pseudoglyptograptus*). *Geologiska Föreningens i Stockholm Förhandlingar,* **94**, 271–280.

—— 1974. A new monograptid genus and the origins of the main monograptid genera. *Special Papers in Palaeontology,* **13**, 141–147, pl. 9.

—— 1988. Graptolite faunas at the base of the Silurian. *Bulletin of the British Museum (Natural History), Geology Series,* **43**, 345–349.

—— BRUSSA, E., TORO, B. and ORTEGA, G. 1996. Ordovician and Silurian graptolite assemblages from Cerro del Fuerte, San Juan Province, Argentina. *Geological Journal,* **31**, 101–122.

—— and KOREN', T. N. 1974. Virgellar meshworks and sicular spinosity in Llandovery graptoloids. *Geological Magazine,* **111**, 193–204, pls 1–2.

RIGBY, S. 2000. *Akidograptus ascensus* Davis [sic], 1929. *Atlas of graptolite type specimens.* Folio 1. Palaeontographical Society, London.

RIVA, J. 1988. Graptolites at and below the Ordovician-Silurian boundary on Anticosti Island, Canada. *Bulletin of the British Museum (Natural History), Geology Series,* **43**, 221–237.

RONG JIA-YU, CHEN XU, ZHAN REN-BIN, ZHOU ZHI-QIANG, ZHENG ZHAO-CHANG and WANG YI 2003. Late Ordovician biogeography and Ordovician-Silurian boundary in the Zhusilenhaierhan area, Ejin, western Inner Mongolia. *Acta Palaeontologica Sinica,* **42**, 149–167.

RUEDEMANN, R. 1929. Description of the Rio Trombetas graptolites. 21–25, figs 1–3. *In* MAURY, C. J. Uma zona de Graptolitos do Llandovery inferior no Rio Trombetas, Estado do Pará, Brasil. *Servico Geologico e Mineralogico do Brasil, Monographia,* **7**, 6–45.

RUSSEL, J. C., MELCHIN, M. J. and KOREN', T. N. 2000. Development, taxonomy and phylogenetic relationships of species of *Paraclimacograptus* (Graptoloidea) from the Canadian Arctic and the Southern Urals of Russia. *Journal of Paleontology,* **74**, 84–91.

SCHAUER, M. 1971. Biostratigraphie und Taxionomie der Graptolithen des tieferen Silurs unter besonderer Berucksichtigung der tektonischen Deformation. *Freiberger Forschungshefte,* **C273**, Paläontologie, 1–185.

SENNIKOV, N. V. 1976. *Graptolites and lower Silurian stratigraphy of the Gorny Altai.* Nauka Publishing House, Moscow, 274 pp. [In Russian].

SKOGLUND, R. 1963. Uppermost Viruan and lower Harjuan (Ordovician) stratigraphy of Västergötland and lower Harjuan graptolite faunas of central Sweden. *Bulletin of the Geological Institutions of the University of Uppsala,* **42**, 1–55, pls 1–5.

SOBOLEVSKAYA, R. F. 1974. New Ashgill graptolites in the middle flow basin of the Kolyma River. 63–71, pl. 3. *In* OBUT, A. M. (ed.). *Graptolites of the USSR.* Nauka, Siberian Branch, Novosibirsk, 160 pp. [In Russian].

STEIN, V. 1965. Stratigraphische und paläontologische Untersuchungen im Silur des Frankenwaldes. *Neues Jahrbuch für Geologie und Paläontologie, Abhandlungen,* **121**, 111–200, pls 14–15.

ŠTORCH, P. 1983a. The genus *Diplograptus* (Graptolithina) from the lower Silurian of Bohemia. *Věstník Ústředního Ústavu Geologického,* **58**, 159–170, pls 1–4.

—— 1983b. Sufamily Akidograptinae (Graptolithina) from the lowermost Silurian of Bohemia. *Věstník Ústředního Ústavu Geologického,* **58**, 295–299, pls 1–2.

—— 1985. *Orthograptus s. l.* and *Cystograptus* (Graptolithina) from the Bohemian lower Silurian. *Věstník Ústředního Ústavu Geologického,* **60**, 87–99, pls 1–4.

—— 1986. Ordovician-Silurian boundary in the Prague Basin (Barrandian area, Bohemia). *Sbornik Geologických, Ved. Geologie,* **41**, 69–103, pls 1–10.

—— 1988. Earliest Monograptidae (Graptolithina) in the lower Llandovery sequence of the Prague Basin (Bohemia). *Sbornik Geologických, Ved. Paleontologie,* **29**, 9–48, pls 1–12.

—— 1989. Late Ordovician graptolites from the upper part of the Králův Dvůr Formation of the Prague Basin (Barrandian

Bohemia). *Vestník Českého Geologického Ústavu*, **64**, 173–186, pls 1–2.

—— 1994. Graptolite biostratigraphy of the Lower Silurian (Llandovery and Wenlock) of Bohemia. *Geological Journal*, **29**, 137–165.

—— 1996. The basal Silurian *Akidograptus ascensus – Parakidograptus acuminatus* Biozone in peri-Gondwanan Europe: graptolite assemblages, stratigraphical ranges and palaeobiogeography. *Vestník Českého Geologického Ústavu*, **71**, 177–188.

—— —— 2006. Facies development, depositional settings and sequence stratigraphy across the Ordovician-Silurian boundary: a new perspective from the Barrandian area of the Czech Republic. *Geological Journal*, **41**, 163–192.

—— and LOYDELL, D. K. 1996. The Hirnantian graptolites *Normalograptus persculptus* and '*Glyptograptus*' *bohemicus*: stratigraphical consequences of their synonymy. *Palaeontology*, **39**, 869–881.

—— and MASSA, D. 2004. Biostratigraphy, correlation, environmental and biogeographic interpretations of the lower Silurian graptolite faunas of Libya. 237–51. In SALEM, M. J. and OUN, K. M. (eds). *The geology of northwest Libya, Vol. 1, Sedimentary basins of Libya – Second symposium*. Earth Science Society of Libya, Tripoli, 339 pp.

—— and MASSA, D. 2006. Middle Llandovery (Aeronian) graptolites of the western Murzuq Basin and Al Qarqaf Arch region, south-west Libya. *Palaeontology*, **49**, 83–112.

—— and SERPAGLI, E. 1993. Lower Silurian graptolites from southwestern Sardinia. *Bollettino della Società Paleontologica Italiana*, **32**, 3–57.

STRACHAN, I. 1971. A synoptic supplement to 'A monograph of British graptolites by Miss G. L. Elles and Miss E. M. R. Wood'. *Monograph of the Palaeontographical Society*, **125** (529), 1–130.

—— 1997. A bibliographic index of British graptolites (Graptoloidea). Part 2. *Monograph of the Palaeontographical Society*, **151** (603), 41–155.

SUESS, E. 1851. Über böhmische Graptolithen. *Naturwissenschaftliche Abhandlungen*, **4** (4), 87–134, pls 7–9.

TELLER, L. 1969. The Silurian biostratigraphy of Poland based on graptolites. *Acta Geologica Polonica*, **19**, 393–501.

TOGHILL, P. 1968. The graptolite assemblages and zones of the Birkhill Shales (Lower Silurian) at Dobb's Linn. *Palaeontology*, **11**, 654–668.

—— 1970. Highest Ordovician (Hartfell Shales) graptolite faunas from the Moffat area, south Scotland. *Bulletin of the British Museum (Natural History), Geology Series*, **19**, 1–26, pls 1–16.

TOMCZYK, H. 1962. Występowanie form *Rastrites* w dolnym Sylurze gór świętokrzyskich. *Instytut Geologiczny Biuletyn*, **174**, 65–92, pls 1–8.

TÖRNQUIST, S. L. 1897. On the Diplograptidæ and Heteroprionidæ of the Scanian Rastrites-Beds. *Lunds Universitets Årsskrifter*, **33**, 1–24, pls 1–2.

—— 1899. Researches into the Monograptidae of the Scanian Rastrites Beds. *Lunds Universitets Årsskrifter*, **35**, 1–26, pls 1–4.

UNDERWOOD, C. J., DEYNOUX, M. and GHIENNE, J.-F. 1998. High palaeolatitude (Hodh, Mauritania) recovery of graptolite faunas after the Hirnantian (end Ordovician) extinction event. *Palaeogeography, Palaeoclimatology, Palaeoecology*, **142**, 91–105.

URBANEK, A., KOREN', T. N. and MIERZEJEWSKI, P. 1982. The fine structure of the virgular apparatus in *Cystograptus vesiculosus*. *Lethaia*, **15**, 207–228.

VANDENBERG, A. H. M., RICKARDS, R. B. and HOLLOWAY, D. J. 1984. The Ordovician-Silurian boundary at Darraweit Guim, central Victoria. *Alcheringa*, **8**, 1–22.

WÆRN, B. 1948. The Silurian strata of the Kullatorp core. *Bulletin of the Geological Institutions of the University of Upsala*, **32**, 433–474, pl. 26.

WANG XIAO-FENG 1987. Lower Silurian graptolite zonation in the eastern Yangzi (Yangtze) Gorges, China. *Bulletin of the Geological Society of Denmark*, **35**, 231–243.

WILLIAMS, S. H. 1982. Upper Ordovician graptolites from the top Lower Hartfell Shale Formation (*D. clingani* and *P. linearis* zones) near Moffat, southern Scotland. *Transactions of the Royal Society of Edinburgh: Earth Sciences*, **72**, 229–255.

—— 1983. The Ordovician-Silurian boundary graptolite fauna of Dob's Linn, southern Scotland. *Palaeontology*, **26**, 605–639, pl. 66.

—— 1987. Upper Ordovician graptolites from the *D. complanatus* Zone of the Moffat and Girvan districts and their significance for correlation. *Scottish Journal of Geology*, **23**, 65–92.

WOLFART, R., BENDER, F. and STEIN, V. 1968. Stratigraphie und Fauna des Ober-Ordoviziums (Caradoc–Ashgill) und Unter-Silurs (Unter-Llandovery) von Südjordanien. *Geologisches Jahrbuch*, **85**, 517–564.

YANG DA-QUAN 1964. Some lower Silurian graptolites from Anji, northwestern Zhejiang (Chekiang). *Acta Palaeontologica Sinica*, **12**, 628–635, pl. 1.

YU JIAN-HUA, FANG YI-TING and ZHANG DA-LIANG 1988. Lungmachi Formation graptolites from Salangpu of Xixiang, southern Shaanxi. *Acta Palaeontologica Sinica*, **27**, 150–163, pls 1–5.

ZALASIEWICZ, J. 1996. Aeronian (Silurian: Llandovery) graptolites from central Wales. *Geologica et Palaeontologica*, **30**, 1–14.

—— and TUNNICLIFF, S. 1994. Uppermost Ordovician to lower Silurian graptolite biostratigraphy of the Wye Valley, central Wales. *Palaeontology*, **37**, 695–720.

APPENDIX

The material studied for this work is housed in a number of institutions. To enable future workers to track down material of particular taxa easily a list of all specimens examined and their registration numbers is provided below. Note that some specimens bear more than one example of the taxon concerned.

Neodiplograptus africanus (Legrand, 1970): BGS FOR 5369l, 5369t (core BG-14, 30.0 m); BGS FOR 5388 (core BG-14, 35.0 m); GD-NRA 2402-9b, 2402-9f (core BG-4, 4.9 m).

Neodiplograptus apographon (Štorch, 1983a): BGS FOR 5359a, 5359b, 5359c, 5359e, 5359i, 5423b, 5423c, 5425 (core BG-14, 43.4 m); GD-NRA 8402-36e, 8402-36j, 8402-36k (core BG-14, 42.25 m); GD-NRA 8402-35(1)a (core BG-14, 42.6 m); GD-NRA 8402-30(1a), 8402-30(1)d, 8402-30(2)a, 8402-30(3)a (core BG-14, 46·62 m); SM A109359 (Loc. 12).

Neodiplograptus lanceolatus Štorch and Serpagli, 1993: BGS FOR 5359d, 5422a, 5422b (core BG-14, 43.4 m); BGS FOR 5427 (core BG-14, 44.5 m); BGS FOR 5432a (core BG-14, 46.3 m); BGS FOR 5443b (core BG-14, 46.8 m); BGS FOR 5458a (Loc. 1); BGS FOR 5464 (core WS-6, 1398.3 m); BGS FOR 5465a, 5465i, 5465l (core WS-6, 1399.0 m); BGS FOR 5466 (core WS-6, 1399.05 m); BGS FOR 5471b, 5471f (core WS-6, 1399.5 m); GD-NRA 2402-8a, 2402-8b (core BG-4, 9.1 m); GD-NRA 2402-7 (core BG-4, 10.5 m); GD-NRA 2402-6 (core BG-4, 10.77 m); GD-NRA 2402-4 (core BG-4, 13.8 m); GD-NRA 8402-37 (core BG-14, 41.85 m); GD-NRA 402-36a, 8402-36f, 8402-36g (core BG-14, 42.25 m); GD-NRA 8402-35(2)a, 8402-35(2)f, 8402-35(3)a (core BG-14, 42.6 m); GD-NRA 8402-30(5), 8402-30(6) (core BG-14, 46.62 m); NHM QQ 234–236 (Loc. 6); SM A109351 (core WS-6, 1399.8 m).

Neodiplograptus lueningi sp. nov.: BGS FOR 5420a (core BG-14, 39.8 m); BGS FOR 5430d (core BG-14, 455 m); GD-NRA 8402-34a (core BG-14, 43.3 m).

Neodiplograptus modestus (Lapworth, *in* Armstrong, Young and Robertson 1876): GD-NRA 18402/54d (Loc. 7).

Neodiplograptus? opimus sp. nov.: GD-NRA 8402-31 (core BG-14, 46.0 m).

Neodiplograptus parajanus (Štorch, 1983a): BGS FOR 5458c (Loc. 1); BGS FOR 5468 (core WS-6, 1399.3 m); BGS FOR 5471d, 5472a (core WS-6, 1399.5 m); GD-NRA 8402-35(1)b (core BG-14, 42.6 m).

Neodiplograptus shanchongensis (Li, 1984): BGS FOR 5466c (core WS-6, 1399.05 m); GD-NRA 2402-3a (core BG-4, 16 m).

Neodiplograptus sp. 1: BGS FOR 5437a (core BG-14, 46.3 m); GD-NRA 8402-32(1)a (core BG-14, 44.7 m).

Neodiplograptus sp. 2: BGS FOR 5369o (core BG-14, 30.0 m).

Neodiplograptus sp. 3: BGS FOR 5363d, 5367c (core BG-14, 27.5 m); BGS FOR 5368b (core BG-14, 28.5 m); BGS FOR 5361b (core BG-14, 30.0 m); BGS FOR 5457, 5459 (Loc. 1); BGS FOR 5471e (core WS-6, 1399.5 m); GD-NRA 2402-9e, 2402-9g (core BG-4, 4.9 m).

Neodiplograptus sp. 4: NHMQ 6339(9), 6339(10) (Loc. 5).

Neodiplograptus sp. 5: BGS FOR 5464g (core BG-14, 1398.3 m).

Normalograptus ajjeri (Legrand, 1977): BGS FOR 5369k, 5369p, 5369s, 5369u, 5372b (core BG-14, 30.0 m); BGS FOR 5408 (core BG-14, 36.8 m); BGS FOR 5409a, 5410a, 5410c, 5412, 5414a,

5415 (core BG-14, 37.5 m); BGS FOR 5417a (core BG-14, 39.0 m); BGS FOR 5424 (core BG-14, 43.4 m); BGS FOR 5429a (core BG-14, 44.5 m); BGS FOR 5438b, 5441, 5443c (core BG-14, 46.8 m); BGS FOR 5447c, 5453a, 5455 (Loc. 4); BGS FOR 5468g (core WS-6, 1399.3 m); BGS FOR 5469a (core WS-6, 1399.45 m); GD-NRA 2402-9 (core BG-4, 4.9 m); GD-NRA 2402-3b, 2402-3c, 2402-3d (core BG-4, 16 m); GD-NRA 8402-25a (core BG14, 21.5 m); GD-NRA 8402-36b (42.25 m); GD-NRA 8402-35(1)c, 8402-35(2)c, 8402-35(3)b (core BG14, 42.6 m); GD-NRA 8402-34e (core BG-14, 43.3 m); GD-NRA 8402-32(1)e (core BG-14, 44.7 m); GD-NRA 8402-30(1)e, 8402-30(3)b, 8402-30(7)a (core BG-14, 46.62 m); GD-NRA 18402/57a, 18402/57c (Loc. 7); NHM Q 6339(1) (Loc. 5); SM A109354d (Loc. 13).

Normalograptus angustus (Perner, 1895): BGS FOR 5410d (core BG-14, 37.5 m); BGS FOR 5417c (core BG-14, 39.0m); BGS FOR 5420e (core BG-14, 39.8 m); BGS FOR 5359f (core BG-14, 43.4 m); BGS FOR 5438c (core BG-14, 46.8 m); GD-NRA 8402-37a (core BG-14, 41.85 m); GD-NRA 8402-36h (core BG-14, 42.25 m); GD-NRA 8402-35(2)b, 8402-35(2)e (core BG-14, 42.6 m); GD-NRA 8402-30(1)f (core BG-14, 46.62 m).

Normalograptus bifurcatus sp. nov.: GD-NRA 2402-3a (core BG-4, 16.0 m); NHM Q 6336–6341 (Loc. 5).

Normalograptus cortoghianensis (Štorch and Serpagli, 1993): BGS FOR 5400b (core BG-14, 36.0 m).

Normalograptus larini Koren' and Melchin, 2000: BGS FOR 5407b (core BG-14, 36.8 m).

Normalograptus medius (Törnquist, 1897): BGS FOR 5369w, 5370 (core BG-14, 30.0 m); BGS FOR 5380c (core BG-14, 32.5 m); BGS FOR 5421b (core BG-14, 42.5 m); BGS FOR 5430f (core BG-14, 45.5 m); BGS FOR 5439, 5443a (core BG-14, 46.8 m); BGS FOR 5464h (core WS-6, 1398.3 m); BGS FOR 5468f (core WS-6, 1399.3 m); BGS FOR 5471c (core WS-6, 1399.5 m); GD-NRA 2402-9a (core BG-4, 4.9 m); GD-NRA 8402-31(2)a (core BG-14, 46.0 m); GD-NRA 8402-30(4)a (core BG-14, 46.62 m); GD-NRA 18402/57g, 18402/57g (Loc. 7).

Normalograptus mirnyensis (Obut and Sobolevskaya, 1967): BGS FOR 5409b (core BG-14, 37.5 m); BGS FOR 5416a, 5417b, 5418a (core BG-14, 39.0 m); BGS FOR 5419a, 5420b, 5420g, 5420i (core BG-14, 39.8 m); BGS FOR 5421a (core BG-14, 42.5 m); GD-NRA 18402/48(1)a, 18402/48(1)b, 18402/48(2)a (Loc. 7); SM A109356 (Loc. 13).

Normalograptus normalis (Lapworth, 1877): BGS FOR 5378a (core BG-14, 32.5 m); BGS FOR 5360g, 5397b, 5400a (core BG-14, 36.0 m); GD-NRA 8402-25a (core BG-14, 21.5 m); GD-NRA 18402/57b, 18402/57f (Loc. 7); SM A109354c, 109356 (Loc. 13).

Normalograptus cf. *normalis* (Lapworth, 1877): BGS FOR 5395b (core BG-14, 36.0 m); BGS FOR 5407a (core BG-14, 36.8 m); BGS FOR 5359g–h, 5423a (core BG-14, 43.4 m); BGS FOR 5428a (core BG-14, 44.5 m); GD-NRA 8402-37b (core BG-14, 41.85 m); GD-NRA 8402-36d, 8402-36i (core BG-14, 42.25 m); 8402-35(3)c, 8402-35(3)f (core BG-14, 42.6 m); GD-NRA 8402-32(1)b (core BG-14, 44.7 m); 8402-30(5)b (core BG-14, 46.62 m).

Normalograptus parvulus (H. Lapworth, 1900): BGS FOR 5385b, 5385d, 5386, 5393a (core BG-14, 35.0 m); BGS FOR 5414b (core BG-14, 37.5 m); BGS FOR 5419b, 5419c, 5420c, 5420d, 5420f, 5420h (core BG-14, 39.8 m); BGS FOR 5430a, 5430b, 5430e (core BG-14, 45.5 m); BGS FOR 5435 (core BG-14, 46.3 m); BGS FOR 5438d (core BG-14, 46.8 m); BGS FOR 5446 (Loc. 9); BGS FOR 5447a–b, 5447d, 5448, 5449a, 5449c, 5450a–b, 5452, 5453b, 5454 (Loc. 4); BGS FOR 5456 (Loc. 1); BGS FOR 5465 (core WS-6, 1399.0 m); BGS FOR 5471, 5472b (core WS-6, 1399.5 m); GD-NRA 8402-37 (core BG-14, 41.85 m); GD-NRA 8402-36c (core BG-14, 42.25 m); GD-NRA 8402-35(2)d, 8402-35(3)d (core BG-14, 42.6 m); GD-NRA 8402-34b–d, 8402-34f–g (core BG-14, 43.3 m); GD-NRA 8402-32f (core BG-14, 44.7 m); GD-NRA 8402-31(1), 8402-31(2)c (core BG-14, 46.0 m); GD-NRA 8402-30(1)b–c, 8402-30(3)c (core BG-14, 46.62 m); GD-NRA 18402/48(2)b, 18402/51, 18402/53, 18402/58 (Loc. 7); NHM Q 6337(1), 6339(7)–(8), 6340(1), 6341(2) (Loc. 5); NHM QQ 235(2)–(3) (Loc. 6); SM A109352 (core WS-6, 1399.8 m); SM A109356 (Loc. 13); SM A109358 (Loc. 12); SM A109364, 109365 (core JF-1, 3494' (*c.* 1065 m)).

Normalograptus persculptus (Elles and Wood, 1907): NHM Q 6339(6) (Loc. 5); SM A109366 (core JF-1, 3497' (*c.* 1066 m)).

Normalograptus rectangularis (McCoy, 1850): BGS FOR 5368c (core BG-14, 28.5 m); BGS FOR 5378b, 5380b, 5380d (core BG-14, 32.5 m); BGS FOR 5353a–c (core BG-14, 33.5 m); BGS FOR 5461–2, 5463b (Loc. 10); GD-NRA 18402/54a (Loc. 7); SM A109355–109356 (Loc. 13).

Normalograptus targuii Legrand, 2001: BGS FOR 5384a, 5393b (core BG-14, 35.0 m); BGS FOR 5360d–f, 5360h, 5396a–b, 5397a, 5399a–h (core BG-14, 36.0 m); BGS FOR 5401a, 5404, 5406 (core BG-14, 36.8 m); GD-NRA 18402/49b (Loc. 7).

Normalograptus transgrediens (Wærn, 1948): BGS FOR 5368a (core BG-14, 28.5 m).

Normalograptus trifilis (Manck, 1923): BGS FOR 5401b (core BG-14, 36.8 m).

Normalograptus sp.: BGS FOR 5369m (core BG-14, 30.0 m).

Paraclimacograptus libycus (Desio, 1940): NHM QQ 230–233 (Loc. 6).

Paraclimacograptus obesus (Churkin and Carter, 1970): BGS FOR 5375a, 5375c–d (core BG-14, 30.0 m).

Paraclimacograptus sp.: BGS FOR 5371a (core BG-14, 30.0 m).

Metaclimacograptus hughesi (Nicholson, 1869): BGS FOR 5363b–c, 5367a–b (core BG-14, 27.5 m); GD-NRA 18402/55(1)–(2) (Loc. 7).

Sudburigraptus illustris (Koren' and Mikhaylova, 1980): BGS FOR 5371b, 5374 (core BG-14, 30.0 m); SM A109362 (core JF-1, 3490' (*c.* 1064 m)).

Sudburigraptus sp.: BGS FOR 5369q, 5372a, 5373a (core BG-14, 30.0 m).

Cystograptus vesiculosus (Nicholson, 1868*b*): GD-NRA 18402/54 (Loc. 7).

Akidograptus ascensus Davies, 1929: BGS FOR 5464e–f (core WS-6, 1398.3 m); BGS FOR 5465 (core WS-6, 1399.0 m); BGS FOR 5468h (core WS-6, 1399.3 m); BGS FOR 5470 (core WS-6, 1399.45 m); SM A109352 (core WS-6, 1399.8 m).

Parakidograptus acuminatus (Nicholson, 1867): BGS FOR 5360a–b, 5395a, 5396c (core BG-14, 36.0 m); SM A109353–109354, 109356 (Loc. 13).

Rhaphidograptus toernquisti (Elles and Wood, 1906): GD-NRA 18402/54f–g (Loc. 7).

Dimorphograptus confertus (Nicholson, 1868*a*): BGS FOR 5361a (core BG-14, 30.0 m).

Atavograptus atavus (Jones, 1909): BGS FOR 5363a, 5364, 5366 (core BG-14, 27.5 m); BGS FOR 5369a, 5369c–d, 5369f–i (core BG-14, 30.0 m); GD-NRA 18402/54b, 18402/54e, 18402/57 (Loc. 7).

Huttagraptus acinaces (Törnquist, 1899): BGS FOR 5369b, 5473 (core BG-14, 30.0 m), BGS FOR 5463 (Loc. 10).